NATIONAL ACADEMIES
Sciences
Engineering
Medicine

NATIONAL ACADEMIES PRESS
Washington, DC

Cyber Hard Problems

Focused Steps Toward a Resilient Digital Future

Committee on Cyber Hard Problems

Computer Science and Telecommunications Board

Division on Engineering and Physical Sciences

Consensus Study Report

NATIONAL ACADEMIES PRESS 500 Fifth Street, NW Washington, DC 20001

This activity was supported by the Office of the National Cyber Director (ONCD) with the assistance of the National Science Foundation under grant CNS-1933974. Any opinions, findings, conclusions, or recommendations expressed in this publication do not necessarily reflect the views of any organization or agency that provided support for the project.

Any opinions, findings, conclusions, or recommendations expressed in this publication do not reflect the views of ONCD.

International Standard Book Number-13: 978-0-309-73489-9
International Standard Book Number-10: 0-309-73489-4
Digital Object Identifier: https://doi.org/10.17226/29056

This publication is available from the National Academies Press, 500 Fifth Street, NW, Keck 360, Washington, DC 20001; (800) 624-6242 or (202) 334-3313; http://www.nap.edu.

Copyright 2025 by the National Academy of Sciences. National Academies of Sciences, Engineering, and Medicine and National Academies Press and the graphical logos for each are all trademarks of the National Academy of Sciences. All rights reserved.

Printed in the United States of America.

Suggested citation: National Academies of Sciences, Engineering, and Medicine. 2025. *Cyber Hard Problems: Focused Steps Toward a Resilient Digital Future*. Washington, DC: The National Academies Press. https://doi.org/10.17226/29056.

The **National Academy of Sciences** was established in 1863 by an Act of Congress, signed by President Lincoln, as a private, nongovernmental institution to advise the nation on issues related to science and technology. Members are elected by their peers for outstanding contributions to research. Dr. Marcia McNutt is president.

The **National Academy of Engineering** was established in 1964 under the charter of the National Academy of Sciences to bring the practices of engineering to advising the nation. Members are elected by their peers for extraordinary contributions to engineering. Dr. John L. Anderson is president.

The **National Academy of Medicine** (formerly the Institute of Medicine) was established in 1970 under the charter of the National Academy of Sciences to advise the nation on medical and health issues. Members are elected by their peers for distinguished contributions to medicine and health. Dr. Victor J. Dzau is president.

The three Academies work together as the **National Academies of Sciences, Engineering, and Medicine** to provide independent, objective analysis and advice to the nation and conduct other activities to solve complex problems and inform public policy decisions. The National Academies also encourage education and research, recognize outstanding contributions to knowledge, and increase public understanding in matters of science, engineering, and medicine.

Learn more about the National Academies of Sciences, Engineering, and Medicine at **www.nationalacademies.org**.

Consensus Study Reports published by the National Academies of Sciences, Engineering, and Medicine document the evidence-based consensus on the study's statement of task by an authoring committee of experts. Reports typically include findings, conclusions, and recommendations based on information gathered by the committee and the committee's deliberations. Each report has been subjected to a rigorous and independent peer-review process and it represents the position of the National Academies on the statement of task.

Proceedings published by the National Academies of Sciences, Engineering, and Medicine chronicle the presentations and discussions at a workshop, symposium, or other event convened by the National Academies. The statements and opinions contained in proceedings are those of the participants and are not endorsed by other participants, the planning committee, or the National Academies.

Rapid Expert Consultations published by the National Academies of Sciences, Engineering, and Medicine are authored by subject-matter experts on narrowly focused topics that can be supported by a body of evidence. The discussions contained in rapid expert consultations are considered those of the authors and do not contain policy recommendations. Rapid expert consultations are reviewed by the institution before release.

For information about other products and activities of the National Academies, please visit www.nationalacademies.org/about/whatwedo.

COMMITTEE ON CYBER HARD PROBLEMS

JOHN MANFERDELLI (NAE), Datica Research, *Chair*
HYRUM ANDERSON, Cisco
JOSIAH DYKSTRA, Trail of Bits
PAUL ENGLAND (NAE), Datica Research
MARITZA JOHNSON, Good Research
ANGELOS D. KEROMYTIS, Georgia Institute of Technology
WENDY NATHER, 1Password
STEFAN SAVAGE (NAE), University of California, San Diego
WILLIAM L. SCHERLIS, Carnegie Mellon University
MARK SEIDEN, Internet Archive
WINDOW SNYDER, Thistle Technologies
MARY ELLEN ZURKO, MIT Lincoln Laboratory

Study Staff

THƠ H. NGUYỄN, Senior Program Officer, Study Director
JON K. EISENBERG, Senior Board Director
SHENAE A. BRADLEY, Administrative Coordinator

COMPUTER SCIENCE AND TELECOMMUNICATIONS BOARD

LAURA M. HAAS (NAE), University of Massachusetts Amherst, *Chair*
DAVID DANKS, University of California, San Diego
CHARLES ISBELL, University of Wisconsin–Madison
ECE KAMAR, Microsoft Research Redmond
JAMES F. KUROSE (NAE), University of Massachusetts Amherst
DAVID LUEBKE, NVIDIA Corporation
DAWN C. MEYERRIECKS, The MITRE Corporation
WILLIAM L. SCHERLIS, Carnegie Mellon University
HENNING SCHULZRINNE, Columbia University
NAMBIRAJAN SESHADRI (NAE), University of California, San Diego
KENNETH E. WASHINGTON (NAE), Medtronic, Inc.

Staff

JON K. EISENBERG, Senior Board Director
SHENAE A. BRADLEY, Administrative Assistant
THƠ H. NGUYỄN, Senior Program Officer
GABRIELLE M. RISICA, Program Officer
AARYA SHRESTHA, Senior Financial Business Partner
NNEKA UDEAGBALA, Associate Program Officer

Reviewers

This Consensus Study Report was reviewed in draft form by individuals chosen for their diverse perspectives and technical expertise. The purpose of this independent review is to provide candid and critical comments that will assist the National Academies of Sciences, Engineering, and Medicine in making each published report as sound as possible and to ensure that it meets the institutional standards for quality, objectivity, evidence, and responsiveness to the study charge. The review comments and draft manuscript remain confidential to protect the integrity of the deliberative process.

We thank the following individuals for their review of this report:

BOB BLAKLEY, Mimic Networks, Inc.
L. JEAN CAMP, Indiana University
VINTON G. CERF (NAS/NAE), Google, LLC
JUAN E. GILBERT, University of Florida
JAMES R. GOSLER (NAE), Johns Hopkins University Applied Physics Laboratory
JOHN CHRIS INGLIS, U.S. Naval Academy
PAUL C. KOCHER (NAE), Independent Researcher
SUSAN LANDAU, Tufts University
CARL E. LANDWEHR, University of Michigan
EUGENE H. SPAFFORD, Purdue University
FLORIAN TRAMÈR, ETH Zürich

Although the reviewers listed above provided many constructive comments and suggestions, they were not asked to endorse the conclusions or recommendations of this report nor did they see the final draft before its release. The review of this report was overseen by **WILLIAM H. PRESS (NAS)**, The University of Texas at Austin, and **STEVEN M. BELLOVIN (NAE)**, Columbia University. They were responsible for making certain that an independent examination of this report was carried out in accordance with the standards of the National Academies and that all review comments were carefully considered. Responsibility for the final content rests entirely with the authoring committee and the National Academies.

Acknowledgments

The committee is grateful for the many experts who generously contributed their time and insight to make this study possible.

We extend our gratitude to the broader community for their engagement with this project. Finally, we appreciate the collaborative efforts of every member of the staff team.

Contents

PREFACE   xiii

SUMMARY   1

1 INTRODUCTION   9
Context, 9
Process for Evaluating and Deliberating the Cyber Hard Problems, 11
Hard Problems from the 1995 and 2005 InfoSec Research Council Reports, 12
Progress Over the Past 20 Years, 15
The Current Cyber Landscape, 16

2 KEY CONSIDERATIONS FOR CYBER RESILIENCY   19
Overarching Considerations, 19
Considerations for Engineering Resilient Cyber Systems, 21
Considerations for Complexity, 25

3 CYBER HARD PROBLEMS   28
Cyber Hard Problem 1: Risk Assessment and Trust, 28
Cyber Hard Problem 2: Secure Development, 30
Cyber Hard Problem 3: System Composition, 33
Cyber Hard Problem 4: Supply Chain, 34
Cyber Hard Problem 5: Policy Establishing Appropriate Economic Incentives, 36
Cyber Hard Problem 6: Human–System Interactions, 37

  Cyber Hard Problem 7: Information Provenance, Social Media, and
    Disinformation, 39
  Cyber Hard Problem 8: Cyber-Physical Systems and Operational
    Technology, 41
  Cyber Hard Problem 9: Artificial Intelligence and Emerging Capabilities, 42
  Cyber Hard Problem 10: Operational Security, 44

4 THE PRODUCER PERSPECTIVE   47
  Functional Cyber Hard Problems, 48
  Operational Cyber Hard Problems, 68
  New Technology Cyber Hard Problems, 80
  Policy Cyber Hard Problems, 89

5 TOWARD COMMUNITY COORDINATION AND PROGRESS   101
  Understanding and Measuring Progress, 102
  Informing Research Investments and Policy Actions, 105

APPENDIXES

A	Statement of Task	109
B	Briefings to the Committee	110
C	Committee Member Biographical Information	112
D	Glossary	118

Preface

Cyber systems are a critical component of society today. New cyber capabilities are continuing to emerge and advance at a dizzying pace, leading to a massive increase in both complexity and ubiquity. The difficulty of building, implementing, and maintaining resilient cyber systems similarly scales. Concurrently, societal factors such as incentives, competition, and geopolitics are making these challenges hard to grasp, let alone tackle. The Office of the National Cyber Director sponsored this study with the assistance of the National Science Foundation in September 2023. The National Academies of Sciences, Engineering, and Medicine convened the Committee on Cyber Hard Problems to conduct a consensus study to create a current list of "cyber hard problems." This effort builds on cyber hard problem lists developed in 1995 and 2005 under the auspices of the federal InfoSec Research Council. The full statement of task is provided in Appendix A.

In distilling and articulating a list of the hard problems that challenge our ability to build high-performing, reliable, and secure cyber systems, the goal of this report is to motivate community action toward addressing them. The list of hard problems and accompanying analyses can serve as a reference to develop research agendas, inform public and private investments, and catalyze new collaborations.

This study leverages the National Academies' extensive work in cybersecurity (including the Forum on Cyber Resilience), national security, and computing and its societal impacts. The committee first met in person in February 2024 and conducted approximately bi-weekly information-gathering sessions through September 2024. It heard from a wide range of actors and stakeholders, including cybersecurity researchers and practitioners, economists, policy experts, and industry operators and users of cyber systems and infrastructure (see Appendix B). These inputs provided the committee with a deeper understanding of cyber technologies, applications, and cyber resilience challenges across multiple sectors.

The report targets an audience consisting of policy makers who need a comprehensive background in the technical issues affecting cybersecurity. The report should also be of interest to researchers, research program managers, engineers, product planners and producers of cyber systems generally, and users ("consumers") of cyber systems. The committee hopes the report illustrates the complex, interconnected set of effects that determine the security and resilience of cyber systems.

Summary

Cyber (computing and communication) technologies underpin every facet of the U.S. economy, are nearly ubiquitous in daily life, and critical for national security. Cyber and cyber-enabled systems are rapidly growing in both complexity and scale, and—despite significant progress—are outpacing the capacity to keep them safe, secure, and resilient to disruptions.[1] Cyber resilience challenges arise from unintentional technical and operational flaws as well as deliberate misuse. Many resist solutions to these challenges because of their technical difficulty, while others resist them owing to intertwined technical, human, business, and policy factors.

Some cyber problems are well defined, and progress toward their solution would significantly improve the safety and resiliency of cyber and cyber-enabled systems. This set of problems, called "cyber hard problems" in this report, stands in contrast to the many other computing and communication problems whose solutions would be beneficial in some way but would not improve resiliency in a meaningful way (e.g., new encryption algorithms), as well as problems whose solutions could have a transformative impact on cyber resiliency but lie outside of the cyber realm (e.g., the geopolitics driving ransomware attacks).

Cyber hard problems are frequently caused or sustained by human or societal factors and misaligned incentives. These in turn are exacerbated by the continuing tremendous growth in the production and use of cyber technologies and their resulting near ubiquity in societally important systems and institutions.

Another contributor to hard problems is the difficulty of measuring cyber resilience or how a particular capability or solution improves it. The resulting failure to establish

---

[1] The terms cyber, cyber resiliency, and cyber-enabled are given more precise definitions in the Glossary (see Appendix D).

incentives hinders the prioritization of investment, research, development, and deployment of new capabilities—often leading to systems that are designed, implemented, deployed, and operated with insufficient cyber resilience. Additionally, the rapid pace of technological advancement and societal adoption of cyber technologies means that policy development often greatly lags behind technology developments.

Similar considerations have prompted several past efforts to develop a cyber hard problems list. In developing a new list, the committee explored critical areas and dimensions of the cyber ecosystem, such as technical development, operations, practices, human–machine interactions, policies and regulations, and incentives.

CREATING A NEW CYBER HARD PROBLEMS LIST

The now-dormant InfoSec Research Council was established by federal agencies sponsoring cybersecurity research. It sponsored studies published in 1995 and 2005[2] that each produced a list of cyber hard problems largely focused on unsolved technical and research problems for which progress toward solutions would have a significant impact on the practical security of cyber systems.

Many of the hard problems listed in those reports remain unsolved, either in theory or practice. In the subsequent two decades, new cyber hard problems have emerged because of dramatic changes affecting cyber resiliency. The most salient factors are as follows:

- The vast increase in the use and adoption of cyber systems by diverse consumers—including individuals, firms, governments, and other organizations—and the attendant interests of major players (approximating those of entire countries).
- The vast increase in the complexity of hardware and software comprising cyber systems; the scale and reach of cyber systems; the scale, scope, and effectiveness of sophisticated cyber attackers; and widespread sharing of data among commercial providers.
- Globalization and diversification of supply chains.
- The rise and internationalization of cloud computing.

---

[2] The 1995 InfoSec Research Council (IRC) *Hard Problems* is not easily found, but the problems themselves are available in Appendix A, "Retrospective on the Original Hard Problem List," of the 2005 *Hard Problems List* report. See IRC, 2005, *Hard Problem List*, November, https://www.nitrd.gov/documents/cybersecurity/documents/IRC_Hard_Problem_List.pdf.

- The vast increase in and pervasive use of cyber-physical systems—such as critical civil infrastructure (power, water, pipelines, etc.), medical systems, and operational technologies for manufacturing, plant operations, civil infrastructure, and national security. This growth of critical cyber-physical systems means that when failures occur, lack of resiliency results in extended outages of business-, safety-, and life-critical systems.
- Widespread adoption of social media and other cyber-mediated influences on opinions and actions.
- Rapid advances and growing adoption of machine learning and other artificial intelligence (AI) technologies.
- Increasing growth of autonomous and semi-autonomous systems.
- Growth in the number of well-resourced, motivated, and sophisticated state and non-state attackers.
- Adoption of AI and other advanced technologies by attackers.

Compounding these challenges are persistent market and policy shortcomings that have failed to incentivize responses that are adequate for meeting society's needs. Thus, a broader analytical lens is needed that acknowledges both the technical and systemic nature of cyber vulnerabilities and accounts for the institutional barriers that impede robust resilience measures.

Cyber hard problems can be approached from the following two perspectives: (1) the key attributes that, if satisfactorily addressed, would enhance cybersecurity and resiliency of cyber systems and (2) the key considerations for developing cyber systems that, if satisfactory progress is made, would enhance the resilience of the resulting system. Another way to view these two perspectives is to see the first as hard problems from the perspective of "consumers" (adopters and users), and the second as hard problems from the perspective of "producers" (vendors and developers). The consumer list is what this report calls the new cyber hard problems list, while the producer list reflects the perspective of those responsible for building cyber and cyber-enabled systems. The two lists overlap significantly, but the different perspectives are nonetheless valuable in understanding the structure of the problems.

CYBERSECURITY AND CYBER RESILIENCE

"Cybersecurity" refers to the customary security dimensions of confidentiality, integrity, and availability of information in accordance with explicit security policies, along with several other key attributes. A secure system (1) does not interfere with the correct operation of "adjacent systems" in an organization or on a network;[3] (2) operates with ease of use and appropriate transparency for human operators and users; (3) is resilient to failures both within the system and on the part of the operators and users, repairing itself or gracefully degrading rather than failing entirely; and (4) is resistant to attacks from knowledgeable adversaries, but when compromises do occur, the affected parts of the system should minimally impair operation of the rest of the system.

This latter characteristic of resilience to failures and attacks—including avoiding cascading failures—is particularly important as systems scale up and become more highly interconnected. Achieving resilience can be challenging because adversaries may have more knowledge of the internals of a system than its own operators and users, who are bound by limitations on their organizational roles, confidentiality provisions of license agreements, and their abilities to collaborate that adversaries are free to ignore. Moreover, adversaries, in many circumstances, may have access to the communications infrastructure in the guise of legitimate operators and users, and indeed, as recently reported, at levels of access that go beyond even that of systems operators. The committee also considered impacts on individuals and society though social media and similar channels as another facet of cyber resilience.

THE 2025 CYBER HARD PROBLEMS

The following cyber hard problems (numbered CHP1–CHP10) identify the areas of focus identified as most significant and challenging by the committee, where advances in technology, practice, or policy would make a measurable difference.

- *CHP1—Risk assessment and trust.* How can all aspects of cyber risk be better assessed, including system vulnerability and attack surfaces, operational consequences of failures, resilience to attack, and characteristics of threats? Can incentive configuration and risk assessment capacity be improved to enable more reliable, informed choices in consequential applications, particularly where multiple stakeholders are needed to support progress on complex systems? Such progress would create new opportunities to drive efficiencies and deliver novel—and consequential—capabilities in the full range of applications for computation-based systems.

---

[3] This includes cascading failures, for example.

- *CHP2—Secure development.* How does one reliably engineer systems that are secure "out of the box" and safely evolve them in response to changing needs? The goal of building in security has long been asserted, but there are challenges in accomplishing this with respect to incentives and tools, techniques, and practices. Incentives are a challenge because it is difficult to measure if certain practices are yielding security improvements and determine if the cost or risk in adopting those practices is justified. Tools, techniques, and practices can be discounted because developers often insist on achieving significant improvements while not impairing productivity or system performance. However, there is evidence that practices are emerging that have negative cost, in that they increase developer productivity and create means to support rapid system evolution with continuous assurance.
- *CHP3—Secure composition.* What are the technical principles for securely integrating larger-scale systems from diverse components and services? The goal of secure composition is to enable reliance on separately made security judgments regarding individual system elements—both components and services—to support efficient judgments regarding the composite system. This is an issue almost universally faced due to the success of software libraries, frameworks, and reuse generally. For Web applications, for example, component libraries include millions of open-source and vendor components from which developers can draw.
- *CHP4—Supply chain.* How can one securely develop and manage large software and hardware systems engineering projects when there are diverse sources of components and services? A significant challenge for system designers, implementers, and sustainers is how to confidently assemble such systems when supply chains are complex and, very often, opaque due to trade secrecy and other considerations. There are technical principles, such as architecting for least privilege (which includes zero trust), and there are also policy principles, such as offering some degree of "translucency" regarding vendor components and services.
- *CHP5—Policy establishing appropriate economic incentives.* How can liability and accountability be allocated in a way that both encourages higher levels of security and also promotes rapid innovation? It is evident that considerations of cost and time-to-market often dominate security-focused engineering practices. Indeed, the difficulty of measuring levels of security can remove incentives to enhance security, since there may be no easy way to measure and thus reward the outcome. Process compliance and other surrogates are helpful but not sufficient in the face of modern attackers. A combination of technical

and policy interventions could create incentives to address this "measurement conundrum" and enable producers to be rewarded for enhanced levels of security.

- *CHP6—Human–system interactions.* How can systems be designed in ways that reduce the extent to which attackers are able exploit human behavior to gain access to systems? Phishing and social engineering attacks are currently the dominant means of adversary access to all categories of systems, from mobile devices to governmental systems. Social science has shown that there is not an immutable trade-off of security and usability, and that with good engineering informed by social science, systems can be both secure and usable by operators and end users. How can systems be designed to support effective interactions with humans to support security-related activities ranging from authentication to operational attack response?

- *CHP7—Information provenance, social media, and disinformation.* How can social media support free speech and exchange of ideas while also protecting the safety and privacy of users and keeping them alert to deep fakes and false information. There is an ongoing war of attrition between developers of deep fake media, including fraudulent images, videos, and text, and those attempting to detect and flag such fakes—with modern AI technology having strong roles on both sides. Nation states may launch campaigns that exploit access to online user profiles to accomplish precision targeting. There are also emerging technical approaches to watermark (or otherwise attribute) media creations to facilitate tracking of information provenance.

- *CHP8—Cyber-physical systems and operational technology.* How can one better secure the Internet of Things and operational technology devices that are the central nervous system for manufacturing, civil infrastructure, and transportation, as well as for the many systems that operate in homes and offices? Many operational technology devices were designed on the assumption, now mostly false, that they would not be connected to the Internet and thus not exposed to cyber threats. Software and firmware updates are also challenging for many of these systems. Making matters worse is the complexity and modeling difficulty and uptime requirements due to their role in managing real-time operations of physical systems.

- *CHP9—AI as an emerging capability.* Challenges include the use of AI in offensive cyberoperations, and more broadly, the use of AI *in*—or *as*—mainstream software. Indeed, many of the challenges associated with the fact that AI models are increasingly being incorporated into software systems are implicitly addressed in other problem statements. There are, however, challenges that are unique to modern AI systems. What are the best techniques to model and analyze modern AI systems to understand the kinds of security-related weaknesses and vulnerabilities that are present? The opaque manner in which models learn from training data and subsequently transform input data into behaviors creates huge challenges for modeling and analysis, not just for functional behaviors but also for aspects ranging from supply chain (e.g., possible effects of poisoned training data) to runtime (e.g., evading guardrails) security. How can AI models be restructured, augmented, or encapsulated to enhance auditability? What drivers will more fully bring secure software design principles to AI systems when the models exhibit unexpected non-smooth and/or non-deterministic behavior? How can run-time controls of generative models be improved when included in scoped applications with specific safety and security requirements? What will be the new security risks associated with offloading more traditionally human-centric tasks to AI-powered autonomous systems? Additionally, AI capabilities are starting to have a significant impact on cyber operations generally, creating new opportunities (and challenges) for system developers, system operators and defenders, and for red teams.

- *CHP10—Operational security*. How can the resilience of the operational systems—the central nervous systems for larger private and governmental organizations—be enhanced in an environment of active threats and high consequence? All aspects of operational security—prevention, detection, response, and recovery—pose challenges to technical design, test and evaluation, operational security practices, and data management—as well as an understanding of the threat environment, legal requirements, and business considerations. Progress would enable improved organizational capability and productivity through automation, despite the presence of sophisticated threats.

CHALLENGES FROM A "PRODUCER" PERSPECTIVE

As discussed earlier, the committee also considered the challenges preventing the producers of cyber systems from solving the problems that consumers see. These include the technical, policy, and operational principles and procedures that the producer has to consider when building resilient cyber systems.

Key challenges for the producer include the following:

- Functional challenges, which deal with the design of secure interoperable products and infrastructure;
- Operational challenges, which concern the secure operation of an "at scale" infrastructure, including responding to attacks in a resilient manner;
- New technology challenges, which are caused by new and emerging computing paradigms and approaches, and their realization and application; for example, the inclusion of AI models in systems and the use of AI models in cyber operations; and
- Policy challenges, which are missing or misaligned policies that result in unexpected consequences to operations; for example, hiding or requiring isolation of previously relied-on data due to privacy concerns.

* * *

Together, the cyber hard problems list and the list of producer challenges provide a useful view of the range of issues associated with cyber resiliency. All stakeholders in the cyber ecosystem—including research funding agencies, computer technology companies, policy makers, and cybersecurity and other computing researchers—can contribute to making progress by reflecting on the driving factors and attacking the hard problems while considering the producer perspective.

1

Introduction

CONTEXT

This report builds on two previous hard problems studies sponsored by the InfoSec Research Council in 1995 and 2005.[1] A lot has changed since those reports were written, but many of their conclusions still apply now, some with even greater force and effect. Today, computing and communication technologies are near-universally integrated in every aspect of society, vastly improving lives but adding new threats, introducing new technology and uses that far outstrip the classic information security perspective of the past.

The Cybersecurity and Infrastructure Security Agency (CISA) recently issued a report[2] illustrating the increasing importance of cyber resilience. The introduction of that report states as follows:

> Access to electricity, transportation, the internet, and a myriad of other services are of paramount importance to the Nation's societal and economic well-being. Each day, critical infrastructure operations ensure that National Critical Functions (NCFs), which serve as the operational backbone for modern society, are running. The NCFs

---

[1] The 1995 Infosec Research Council (IRC) *Hard Problems* report is not easily found, but the problems themselves are available in Appendix A, "Retrospective on the Original Hard Problem List," of the 2005 *Hard Problem List* report. See IRC, 2005, *Hard Problem List*, November, https://www.nitrd.gov/documents/cybersecurity/documents/IRC_Hard_Problem_List.pdf.

[2] Cybersecurity and Infrastructure Security Agency (CISA), "National Critical Functions: A Vital Framework for Cross-Cutting Risk Analysis," Fact Sheet, https://www.cisa.gov/sites/default/files/publications/factsheet_national-critical-functions_508.pdf, accessed February 6, 2025.

are the functions of government and the private sector so vital to the United States that their disruption, corruption, or dysfunction would have a debilitating effect on security, national economic security, national public health or safety, or any combination thereof.

CISA, through the National Risk Management Center (NRMC), works with government and industry partners to identify and manage risks to the NCFs in a targeted, prioritized, and strategic manner to improve the resilience across the United States critical infrastructure.

> Technological advances and hyperconnectivity have improved critical infrastructure operations and transformed the Nation's 16 critical infrastructure sectors into a complex, interconnected ecosystem. At the same time, the integration of information and operational technologies and the complexity of supply chains has created new vectors through which adversaries can exploit vulnerabilities in assets, systems, and networks that enable America's economic competitiveness and national security. Examples of NCFs include electricity generation that powers homes and businesses, transportation of commodities and people, and access to GPS data for cellular networks. An interruption to one NCF can have cascading consequences across industries and society.[3]

The goal of the 2005 cyber hard problems study was different from the present effort. The study was charged with producing "desirable research topics by identifying a set of key problems from a government perspective and in the context of IRC member missions" to "help guide the research program planning of the IRC member organizations."[4] By contrast, the scope of the present study is wider, expanded to "provide a current list of hard problems in cyber resiliency, building on earlier hard problems lists and expanding the scope from cybersecurity to cyber resilience" to "identify ways that the new list could be used to enhance community-wide coordination of R&D [research and development] activities."[5]

This report benefits from the previous work but does not confine itself to identifying and prioritizing *research* hard problems, although it certainly does that. It also examines the ever more critical role of cyber in society in the context of its actual use (e.g., in NCFs, commerce, health care delivery, new mechanisms such as artificial intelligence [AI], and education). The scale, complexity, and impact of platforms, such social media,

---

[3] Ibid.
[4] IRC, 2005, *Hard Problem List*, p. 5.
[5] The statement of task is reprinted in Appendix A.

that are directly used by the general population dramatically increases the effectiveness of disinformation campaigns, adding a new societal dimension to cyber hard problems.

In addition to identifying research challenges, this report also attempts to provide a framework to assess the effectiveness of solving these new or previously identified cyber safety and reliability solutions within the context of the overall cybersecurity landscape as experienced by users in their everyday lives. The framework addresses such questions as the following: What expectations should consumers have about cybersecurity in a particular use context? What are the residual risks that have not been eliminated by technology? What can they do to better understand the risks and thoughtfully mitigate residual vulnerabilities? How can they know the extent to which prior use of attacked systems has resulted in loss, and if so, what can they do to protect themselves? What responsibilities do other parties have?

The definition of *cyber resilience* in this report encompasses all the components of technology, operations, human performance, and policy, including traditional authentication and access control (to allow permitted operations and to prevent the unwanted sharing or corruption of information) and vulnerabilities caused by malfunctioning hardware, software, infrastructure, operations, people (like phishing), or services.[6] Therefore, the goal is to identify a framework to ensure the confidentiality and integrity of cyber processes and the underlying data as well as the resilience and availability of the critical functions they provide.

PROCESS FOR EVALUATING AND DELIBERATING THE CYBER HARD PROBLEMS

The cyber hard problems the committee focused on are well-defined problems whose solution would significantly improve the safety and resiliency of cyber and cyber-enabled systems. It did not focus on problems that were "interesting" but whose solution would not materially improve the safety and resiliency of cyber and cyber-enabled systems. As a related matter, unlike previous hard problem lists, this study was not restricted to research problems or even "technology problems." The committee also considered hard problems related to policy, incentives, and the deployment and operation and use of cyber systems by humans.

In selecting hard problems, the committee spoke with the sponsor, reviewed prior lists, and evaluated how efforts to solve earlier "hard problems" led (or failed to lead) to identifiable improvements. The committee considered problems identified by committee members, previously identified problems that remain fundamentally unsolved

---

[6] See the Glossary in Appendix D.

(or where there has been insufficient progress toward a solution), and suggestions from leading experts in cybersecurity. Since the selection criteria emphasized impact, it heavily weighed priorities in the National Cybersecurity Strategy,[7] the effect of the barriers created by the hard problems on NCFs,[8] past security incidents, emerging new technology that is likely to become widespread, and problems with foundational effects on critical infrastructure.

Finally, to avoid a long laundry list of problems making the list inactionable, the committee selected only the most significant problems where progress toward solution would have the greatest impact.

HARD PROBLEMS FROM THE 1995 AND 2005 INFOSEC RESEARCH COUNCIL REPORTS

The 2005 InfoSec Research Council cyber hard problems study identified the following eight hard problems:[9]

1. **Global-Scale Identity Management:** Global-scale identification, authentication, access control, authorization, and management of identities and identity information
2. **Insider Threat:** Mitigation of insider threats in cyberspace to an extent comparable to that of mitigation in physical space
3. **Availability of Time-Critical Systems:** Guaranteed availability of information and information services, even in resource-limited, geospatially distributed, on demand (ad hoc) environments
4. **Building Scalable Secure Systems:** Design, construction, verification, and validation of system components and systems ranging from crucial embedded devices to systems composing millions of lines of code
5. **Situational Understanding and Attack Attribution:** Reliable understanding of the status of information systems, including information concerning possible attacks, who or what is responsible for the attack, the extent of the attack, and recommended responses
6. **Information Provenance:** Ability to track the pedigree of information in very large systems that process petabytes of information

---

[7] Office of the National Cyber Director, 2023, "The National Cybersecurity Strategy," The White House, March 2, https://bidenwhitehouse.archives.gov/oncd/national-cybersecurity-strategy.
[8] CISA, "National Critical Functions Set." https://www.cisa.gov/national-critical-functions-set, accessed February 6, 2025.
[9] IRC, 2005, *Hard Problem List*, p. 5.

7. **Security with Privacy:** Technical means for improving information security without sacrificing privacy
8. **Enterprise-Level Security Metrics:** Ability to effectively measure the security of large systems with hundreds to millions of users

The hard problems above are not homogeneous in nature. In some cases, solving the problem calls for research into new security mechanisms (global identity management); in some cases, it calls for new analytical techniques (security metrics); and in other cases, it requires a combination of new mechanisms, operations, and design methodologies (building secure systems). The heterogeneity of the hard problems is already visible in the 1995 InfoSec Research Council report. There, the "Functional Hard Problems of 1995" are summarized here:[10]

1. **Intrusion and Misuse Detection:** Commercial systems are still riddled with false positives and false negatives, especially in high-volume situations such as networking. Years of experience has shown that the general problem of intrusion detection leaves adversaries too much room to maneuver, and that the general approaches to intrusion detection are completely blind to certain classes of attack, such as life-cycle attacks. More research is yet to be done. This remains an unsolved problem.
2. **Intrusion and Misuse Response:** Given progress, and the degree to which response depends on detection, the emphasis of this area has been refocused on insider threat detection and detection of covert channels.
3. **Security of Foreign and Mobile Code:** Although difficult research remains, proof-carrying code and sandboxing represent important advances in limiting the potential negative effects of foreign and mobile code. However, even domestic production of software is being outsourced to firms offshore. Moreover, even at reputable software companies, insiders can be bought to plant malicious code into key products used by the U.S. government.
4. **Controlled Sharing of Sensitive Information:** Progress in digital rights management may ease policy specification challenges by empowering end users to set policies as required on information as it is being created. However, without a foundation of trustworthy enforcement mechanisms for enforcing separation, the value of this will be substantially diminished.
5. **Application Security:** Application security has seen important progress toward intrusion-tolerant applications that are able to function despite flawed

---

[10] The 1995 IRC *Hard Problems* report is not easily found, but the problems themselves are available as Appendix A, "Retrospective on the Original Hard Problem List," in IRC, 2005, *Hard Problem List*.

components. These systems can be designed to be less reliant on an underlying trustworthy computing base (TCB) than traditional applications. However, research remains to make these techniques work in distributed, asynchronous, time-critical environments. One of the most painful lessons has been that there will always be situations where the TCB is critical. So, developing truly trustworthy TCBs is still required for building scalable secure systems.

6. **Denial of Service:** Although research remains, progress has been made toward assuring the availability of information systems against denial-of-service attacks. Technology now exists to mitigate distributed denial-of-service attacks. Moreover, progress from traditional fault tolerance can now help mitigate other denial-of-service attacks. However, this technology is generally only available to large service companies.

7. **Communications Security:** The foundation of secure communications is the cryptography and an infrastructure for managing cryptographic keys. Here, practical solutions are economical and in general use. However, secure communications require authenticating security principals, including people as well as computers and programs. Despite progress in secure communications, authentication, which is a critical aspect of these systems, is seldom done properly.

8. **Security Management Infrastructure:** Although research remains, industry has already begun acquiring emerging security response management technologies. Additional research into security response management requires better situational awareness and attack attribution. The remaining work is captured in the operational security hard problem (CHP10) in Chapter 3.

9. **Information Security for Mobile Warfare:** Both homeland defenders and the military now depend on mobile and secure networked computing, particularly given risks of attack and the need for fire, police, rescue, and recovery personnel, to be able to securely coordinate crisis response via information systems, services, and networks. This is largely an implementation and standards problem and not a "basic technology" problem.

Despite progress, many of the problems identified in 1995 and 2005 remain unsolved as a practical matter and are exacerbated by scale, interconnection, and globalization. There are also many new hard problems that arise from societal, technological, infrastructure, and economic progress since 2005 as well as the increased impact of globalization.

PROGRESS OVER THE PAST 20 YEARS

When the Department of Homeland Security (DHS) developed its first (2009) roadmap for cybersecurity research,[11] it cited and drew directly from the 1995 and 2005 hard problem lists prepared by the InfoSec Research Council. Additionally, according to perspectives offered by participants in the previous cyber hard problems lists, the lists were used by individual program officers to defend funding research on a problem on that list.[12] Solicitations for work with an impact on such a problem was convincingly argued to be worthy of funding. There were also a number of convenings and other activities indirectly influenced by the lists and the process of developing them, including the DHS 2009 *Roadmap for Cybersecurity Research*.[13,14]

Concrete progress has been made on some of the hard problems in the 1995 and 2005 reports. This progress is attributable to a number of factors, including research and development investments in deployable and more usable technology.

Security of communication channels can now be regarded largely as a solved research problem because of excellent, widely accepted, and available cryptographic standards and technology.[15] Still, even today's solutions are marred by failures in implementation and operation, including poor key management, weakness in random number generation, attacks on the certificate infrastructure (to which there have already been technical responses such as certificate pinning), attacks on Internet routing due to misconfiguration, and insecurity in the Internet's Domain Name System. Insider risk also continues to affect the ability to solve this problem.[16]

Global identity management has made progress. There is a practical regime for two-factor authentication, and account recovery is routinely available for many commercial applications like health care. Access control for resources that may be widely shared under specified policy remains a problem—not because new mechanisms are needed, but because current implementations are inadequate (for reasons of safety, scale, interoperability, or simply the effort required to use them) for massive multi-user commercial services.

Denial-of-service attacks are no longer a significant threat to larger service providers and edge providers because of investment in multi-geographical presence, big pools of address space, and technology developed to identify and exclude attackers.

---

[11] Department of Homeland Security (DHS), 2009, *A Roadmap for Cybersecurity Research*, November, https://www.dhs.gov/sites/default/files/publications/CSD-DHS-Cybersecurity-Roadmap_0.pdf.
[12] Carl Landwehr, personal communication, December 16, 2024.
[13] DHS, 2009, *A Roadmap for Cybersecurity Research*.
[14] Tomas Vagoun, personal communication, December 16, 2024.
[15] This includes the ongoing evolution of cryptography in response to new technology development such as quantum computing.
[16] Federal Bureau of Investigation, "John Anthony Walker Sr. Spy Case," https://www.fbi.gov/history/artifacts/john-anthony-walker-jr-spy-case, accessed February 6, 2025.

However, small providers typically do not have access to the technology and do not have the operational scale to achieve the same level of safety.

Significant progress has been made in developing systems that eliminate specific vulnerabilities by using type-safe languages (like Rust), fuzzing techniques, and safer libraries. Other problems remain—for example, some network protocols can still be abused as amplifiers of traffic directed at targets.

Operational security has increased in importance, yet most companies and infrastructure cannot practically recover from catastrophic failures and have not implemented the operational procedures to guarantee they can do so economically, as demonstrated by the Colonial Pipeline attack and several public ransomware attacks. (Some pay ransom as the least expensive solution to recover.) End-of-support and end-of-life offerings from vendors either leave organizations vulnerable or force them to incur additional cost by replacing the affected technology.[17]

Patching remains a challenge. Internet of Things (IoT) devices and embedded systems, unlike most "big information technology [IT]" systems, do not provide a practical mechanism for patching. Even for "big IT," patch testing, incremental deployment, and rollback in the case of a flawed patch is sometimes still aspirational, as demonstrated by the wide consequence of a failed CrowdStrike patch to its detection logic for Windows.[18]

THE CURRENT CYBER LANDSCAPE

In the 1980s, as a precursor of the 1995 Infosec Research Council report, a typical microprocessor (e.g., the Motorola 68000) had 68,000 transistors; by 2000, it was millions of transistors, and by 2024 it is tens of billions of transistors. The 1978 "Tour of Unix"[19,20] described an operating system that had about 10,000 lines of C code, developed by a very small team. A modern operating system now consists of tens of millions of lines of code supplied by many different providers, much of it in binary-only device drivers. In the 1980s, computers ran a few critical applications built by a handful of (often domestic) suppliers, many of whom also supplied the hardware. Modern computers run thousands of applications, sourced from many unidentified providers worldwide. Complexity drives capability, but just as effectively it inhibits the ability to assess security.

---

[17] A. Culafi, 2025, "Zyxel Won't Patch End-of-Life Routers Against Zero-Day Attacks," TechTarget, February 5, https://www.techtarget.com/searchsecurity/news/366618782/Zyxel-wont-patch-end-of-life-routers-against-zero-day-attacks.

[18] Georgetown University, 2024, "Is Global Tech Infrastructure Too Vulnerable? Professor Responds to CrowdStrike, Microsoft Outage," July 25, https://www.georgetown.edu/news/ask-a-professor-crowdstrike-outage.

[19] Wikipedia, "A Commentary on the UNIX Operating System," article, https://en.wikipedia.org/wiki/A_Commentary_on_the_UNIX_Operating_System, accessed February 19, 2025.

[20] K. Thompson, 1978, "Unix Implementation," *Bell System Technical Journal* 57(6):part 2.

In 1995 (and even 2005), it was rare for people to interact with their physicians and medical records on their personal computers; relatively rare to do banking using a personal computer; relatively rare to take a course remotely; and relatively rare to interact on social media, watch a movie, or obtain most news online. Now such virtual interactions are commonplace, accelerated by the COVID-19 pandemic, and cyber access to services far outpaces traditional service delivery modes. Cyber increasingly controls physical infrastructure (e.g., Colonial Pipeline), transportation, electric service (transmission and distribution), automated factories, medical devices, and even water and sewer service. All of these operate on legacy technology that interacts with more recent, complex technology over the Internet. All are vulnerable to attack. Physical infrastructure increasingly employs sensors and actuators to operate cyber-physical systems (CPS), including critical infrastructure, home maintenance, medical devices, autonomous vehicles, and automated manufacturing. These CPS often employ security practices that are 10 or 15 years behind IT systems. They are often bespoke and are largely opaque. Both the hardware and software supply chains for CPS are far more diverse than IT—both specialized and global. This further complicates any attempt to address the cybersecurity and resilience challenges posed by these complex systems.[21,22]

Today, very nearly all residents of middle- and high-income countries have access to broadband, smartphones, and personal computers. This was not the case in 1995 or even 2005. The world's population uses this infrastructure to obtain critical services previously obtained in other ways and to control home and office devices.

A few cloud services, such as Amazon Web Services, Microsoft's Azure, and Google Cloud, host corporate applications at centralized data centers located throughout the world. While the operational safety is probably better than most in-house operations, it is opaque to most customers. In 1998 and even 2005, most IT support was an internal function, and outsourcing was comparatively rare. Today, a large fraction of commercial computing has migrated to cloud infrastructure operated by a small number of providers, making them a potentially critical failure node.

The nature and sophistication of attackers has also changed. Attacks on U.S. systems come from around the globe, from actors that are often funded or otherwise supported covertly by nation states. The astonishing accumulation of personal information available from data brokers and collected from a fusion of advertising and social media has made social engineering attacks much more effective. The advent of Bitcoin

---

[21] National Academies of Sciences, Engineering, and Medicine (NASEM), 2022, *Cybersecurity in Transit Systems*, The National Academies Press.
[22] NASEM, 2020, *Communications, Cyber Resilience, and the Future of the U.S. Electric Power System: Proceedings of a Workshop*, The National Academies Press.

and other cryptocurrencies has provided a relatively safe channel for ransom, extortion, and other illicit payments.[23]

Additionally, the last 20 years has seen the rise of social media and the concomitant rise of globally sourced, globally distributed disinformation with little regulation and even less effective protection against it.

To summarize,

- The scale and diversity of services people use every day is vast, where even a single mobile device can have a software payload drawing on thousands of providers.
- Modern cyber systems are among the most complex systems ever fielded and are poorly understood by those using them, including cybersecurity experts.
- Everything is digitized, and many of the data are broadly shared.
- Cyber technology is a broadly diffused set of monocultures.
- Cyber systems sense and control much of the physical world.
- Cloud systems are critical and run an enormous fraction of workloads and service.

Many new hard problems are inspired by these tectonic changes. Chapter 2 explores some of these cross-cutting drivers making hard problems persistently hard.

---

[23] G. Miliefsky, 2024, "Why Do Hackers Love Cryptocurrency?" *Cyber Defense Magazine*, June 6, https://www.cyberdefensemagazine.com/why-do-hackers-love-cryptocurrency.

2

Key Considerations for Cyber Resiliency

Difficulties in maintaining a resilient cyber ecosystem are hardly ever caused by single-dimensional issues. There are fundamental considerations that existed at the inception of computing technologies and persisted as they grew into the cyber ecosystem of today. These considerations are impossible to capture accurately because they cannot be bound; yet, it is important to understand them in order to have a comprehensive view of why cyber hard problems are persistently hard. In this section, the committee analyzes the overarching considerations as well as two key standalone drivers—engineering of resilient cyber systems and system complexity.

OVERARCHING CONSIDERATIONS

There are overarching considerations that affect each and every cyber hard problem described in this report. Many of their adverse effects were evident and relevant in 1995 and 2005, but new considerations have emerged or have been exacerbated due to scale, globalization, policy, and new technology. With increased usage, many technology problems are affected by human, cultural, political, or economic problems. These important influences were, perhaps, less apparent in earlier lists but are undeniable.

These considerations include the following:

- Computing technology is affected by global-scale economics. Scale affects personal data whose shared—and often uncontrolled—use is vast, and whose impacts on privacy are increasing, sometimes in unexpected ways. Financial incentives for collecting personal information and broadly sharing, selling, and repurposing the information are large and growing.

- The vastly increased complexity and interconnection of cyber systems makes it very difficult (and expensive) to implement resilient designs to prevent cyber failures. The complexity is not only due to the increased size and broad deployment of cyber systems but also to the entanglement and integration of disparate stakeholders, data providers, and technology providers within the scope of a single solution ecosystem. Modern services and apps rely on global multi-party ecosystems with interacting policies (e.g., privacy policies) that operate over many jurisdictions. This scaling and interconnection amplify the need to develop architectures that are resilient to compromises and failures in individual system elements.

- Security metrics sufficient to predict or verify the security properties of a cyber system are non-existent. In the absence of metrics based on measurable external parameters of systems, risk assessment has often fallen to the level of penetration testing because establishing the correct context may be difficult. However, there are far too few skillful "pen testers," and the quality of a test too often seems a measure of the skill and luck of the tester. Systems are so large that complete coverage can be too time consuming (and therefore expensive) to be practical. Additionally, many security attributes are contextual—so not amenable to traditional testing—but instead require analysis and inspection, including of design models.

- The opacity and diversity of components in a single cyber system makes the challenges in risk assessment even more difficult. Dependencies among subsystems in a large system may create never-been-tested timing dependencies that only become evident when a disaster forces the entire collection of systems to be restarted for the first time in its life.

- The increasingly complex cyber-physical systems (CPS) and their central role underpinning critical functions mean that such systems are harder to defend, and compromises can cause large, immediate real-world damage. Integration of sensors and actuators broaden the important surface area that must be assured for CPS, such as critical infrastructure, as well for other CPS that can cause real-world damage. As an example, sensors often have their values checked only periodically for exceptional conditions. Local attacks can be timed to cause irreversible damage before the next time a value is checked.

- Complicated human–systems interaction and societal entanglement can confuse users, even in basic interactions, but follow-on consequences can be subtle and important. A significant challenge for designers is that users—in a hurry to accomplish specific tasks or no longer vigilant against rare anomalies—cannot always be counted on to attend to security concerns.

- The emergence of artificial intelligence (AI) (and statistical models) can increase capability; however, due to the inherent weaknesses and consequent vulnerabilities of neuro-network models, it is extremely difficult (in some cases, impossible) to validate decisions made by AI-enabled systems.
- The emergence of new concentrated targets such as clouds, popular software systems, and mobile devices, whose failure could have extremely broad adverse effects and whose complexity and opacity defy principled risk evaluation and principled risk management.

CONSIDERATIONS FOR ENGINEERING RESILIENT CYBER SYSTEMS

There are fundamentally only two operative methodologies to assure resilience of a cyber system. However, both require a complete understanding of the behavior of the entire engineered cyber system.

1. A guarantee that a delivered system will operate in accordance with a well-tailored specification of security-relevant quality behavior, so that an outside observer can be assured that the operations for which the system is employed, if they meet the specification, are safe. Generally, cyber systems are not warranted to operate in accordance with an enforceable behavior; in fact, cyber systems usually come equipped with extensive liability disclaimers protecting the supplier from claims about any but the most obvious, egregious failures and even then, providing incomplete remedies.
2. The development of an accurate risk assessment of every component of a cyber system as well as their interactions. This requires a comprehensive understanding of the cyber system and the components and services that it encompasses, the environment in which it operates, its users, as well as the principles according to which they are composed into the overall system.

Complete understanding of cyber systems is complicated by the fact that modern cyber systems are among the most complex artifacts created by humans, and current cyber systems are literally millions of times more complex than their predecessors 20 years ago, and too complex to model completely. Engineered systems depend, for predictability, on a profound understanding and specification of component subsystems, extensive testing to guarantee conformance with specifications, and complete understanding of the interaction of these subsystems (as well as their human users) to assure safety or, at a minimum, provide actionable, concrete, specific alerts when things have malfunctioned or are about to malfunction. Systems engineered for critical applications also rely on resilience, which is often provided by redundant independent functioning of sub-functions.

Cyber engineering as a discipline offers a few common ways to deal with complexity.

One mechanism is the isolation of each subsystem coupled with extensive interface specification, often called an application program interface, as well as a detailed understanding of the interaction of that subsystem with other subsystems. Indeed, most processors and platforms also incorporate isolation (enclave) technology to partition specific critical software subsystems from other software running on that platform. This is related to the long-standing software engineering principle of information hiding—and the hardware design principle of process separation. These systems also provide mechanisms for storing secrets for that software and allowing that software to cryptographically authenticate itself when interacting with other isolated components.

A modern implementation of this paradigm goes by the rubric confidential computing and incorporates cryptographic hardware roots of trust. These have already been applied to protect limited high-risk data such as storage and use of cryptographic keys or sensitive communications (e.g., the Signal messaging app). However, this level of partitioning and isolation is a rare exception in most cyber systems. Also, such confidentiality mechanisms are not without trade-offs; for example, an attack on the protocols can compromise availability (secrets cannot be decrypted) or integrity (the secrets can be corrupted).

Furthermore, security dogma insists that people and subsystems should exercise well-identified "segregated duties" and any action taken by such an (isolated) subsystem should enforce the "principle of least privilege,"[1] meaning that the segregated subsystem should only take actions when the requesting principal (a software subsystem or a person) has been authenticated and the action conforms with the access control policies for that action, and entities should only have the access to do their jobs and no more. For example, a bank deposit should only be executed if the depositor to an account has been authenticated, and that depositor might not have the right to do transfers or withdrawals.[2]

However, some cybersecurity practices have visibly improved. Both software and hardware development systems (e.g., GitHub) now include mechanisms to track

---

[1] J.H. Saltzer and M.D. Schroeder, 1975, "The Protection of Information in Computer Systems," *Proceedings of the IEEE* 63(9):1278–1308.

[2] This is a consequence of know your customer (KYC) rules developed to make money laundering more difficult.

provenance of contributions. However, more is needed to ensure the integrity of design and program information—the records need to be immutable, and few systems are designed to keep such a rigorous and robust audit trail.

Modern practice also provides for automated testing of components. Recent progress suggests that a comprehensively curated repository with strong provisioning verification, extensive automated tests, and change tracking dramatically shortens the development time, increases reliability, facilitates understanding, and helps rapid reliable upgrades, including upgrades to fix vulnerabilities. Additionally, many actions in support of secure development can also have the effect of enhancing engineering productivity, both in original development and in evolution. However, cyber systems developers may fail to test or verify the majority of functionality.

Many design systems incorporate automated tools to either satisfy compliance requirements or check for identified flaws. These include type-safe languages and associated compilers, automated design simulation and test, static analysis, fuzzing tools, conventional glitch analysis for hardware, and many other tools. Software systems are often developed with type-safe languages and specification-generated internal checks (e.g., for software fault isolation and control flow integrity) to prevent the exploitation of residual software flaws.

Modern complex systems incorporate extensive monitoring and logging as well as automated tools to analyze collected data to spot problems.

The state of formal verification has advanced and can provide principled protection. Historically, the protection is usually complete only for relatively small systems, but there is a growing population of larger-scale examples.[3] More often, the focus is on more narrow critical properties, in which case greater degrees of scaling can be achieved. An example of this is type-safety in languages, such as Java, Rust, and Typescript, where critical properties "come for free" and, additionally, developers may not even be aware of the security benefits that come along with the increased productivity.

Cryptography has provided a firm underpinning for communication security.

These techniques offer important "partial" solutions to the cross-cutting resilient design problem. However, challenges remain.

- Side-channel attacks are very difficult to identify and often impossible to avoid. Side channels arise from unspecified, but observable, implementation characteristics (like power consumption or timing) that inadvertently disclose information that was thought to be protected.

---

[3] B. Cook, 2024, "An Unexpected Discovery: Automated Reasoning Often Makes Systems More Efficient and Easier to Maintain," *Amazon Web Services Security Blog*, October 17, https://aws.amazon.com/blogs/security/an-unexpected-discovery-automated-reasoning-often-makes-systems-more-efficient-and-easier-to-maintain.

- Isolation and partitioning can offer protection for a relatively small system with well understood and carefully designed components but does not provide complete protection for complex systems with only partially understood or cursorily verified operations. Strict adherence to security-focused architecture designs can avert problems, but the required rigor is seldom employed.
- Systems are often built in an ad hoc way from many existing and widely used subsystems, which are themselves poorly understood. The most visible example is the incorporation of a vast library of open-source components. The web-focused open-source NPM (node package manager) library, for example, has more than 3 million packages available to web developers, along with various package management capabilities. Another common example is that, in the interest of economy, systems are frequently assembled using legacy subsystems that were not even built with then existing security standards.
- The way software contributions enter open-source projects varies widely. Attempts using social engineering to add flawed, exploitable code to a widely used open-source code brought additional attention to the difficulties of assurance in open-source projects.[4]
- Even well-designed systems depend entirely on operational configuration and policy management to operate properly; these systems and procedures are themselves complex and often error prone, and especially difficult to debug.
- Finally, modern cyber systems are heterogeneous, and a complete analysis relies on understanding the design and function of hardware, software, sensors and actuators, communications infrastructure, a bit of mathematics, and any relied-on external services.

Process assurance, including compliance regimes that include third-party verification of identified properties, can help a little but is usually an inadequate substitute for well-informed, engineered resilience based on direct modeling, analysis, and evidence. Process assurance is also often influenced by powerful stakeholders who can help shape rulemaking, placing small providers at a disadvantage in security perception and standards-based acquisition.

As a result, well-designed complex cyber systems often employ active remediation like rapid patching and reliable verifiable recovery as a resilient design crutch to compensate for future discovered flaws.

---

[4] S. Sabin, 2024, "Open-Source Developers Face a Potential Social-Engineering Crisis," *Axios*, April 19, https://www.axios.com/2024/04/19/open-source-software-social-engineering-hacks.

CONSIDERATIONS FOR COMPLEXITY

The previous section analyzed the resilience of small- or medium-scale cyber systems designed by a single supplier. However, the complexity of modern cyber systems, including having been integrated from multiple producers, introduces even greater challenges, including the following:

1. Verification of the function and qualities of acquired subsystem components, including open source.
2. The global hardware supply chain, which includes widely sourced complex components that can include deliberately inserted vulnerabilities.
3. The technical, economic, political, and cultural dynamics that make it difficult, if not impossible, for individual organizations to secure themselves effectively when they are dependent on diverse, complex integrated systems.
4. Relocated risk associated with deploying systems in one of a few large cloud providers whose infrastructure, operations, and even legal jurisdiction is undisclosed. Although most cloud providers offer safer infrastructure than less well-trained and well-practiced users provide for themselves—and they have strong business incentives to deliver high levels of security—they can still be single points of failure that become focused targets of malign nation states and rogue law enforcement and are well shielded from consequential liability by contract and law.
5. The sheer effect of scale that provides capability built on vast compositions of and interdependencies among complex systems.
6. The lack of clear metrics, monitoring, and operational procedures to accurately reflect residual risk.
7. Resilience and recovery (returning to a known-good state or, more precisely, to a well-understood state) from catastrophes is greatly complicated by the complexity and scale of modern cyber systems. This includes data and system configuration as well as hardware and software components. This is partly a technical problem but also part of the operational challenge discussed above.
8. Automated audit and forensics to find sources and effects of compromise are not widely available or widely used.
9. Development of standards for infrastructure that enable safety and assessment is not prescriptive enough to determine liability for a failure.
10. Data provenance integrity is seldom well implemented.
11. Protection of critical data (like PII) is not reliable against attacks from skillful threats, foreign and domestic, outsiders and insiders. Consider the attacks

against the U.S. Office of Personnel Management[5,6] and on unpatched vulnerable infrastructure particularly intended for legal interception of telephone calls at multiple telephone carriers internationally ("Salt Typhoon" or "RedMike").[7]

Complexity is increasingly a factor in limiting the extent to which it is possible to understand a cyber system or CPS. However, proprietary barriers also prevent examination of code, design, hardware, and operations.

Many products have been admirably assessed by normal "market" incentives and mechanism, not cyber. Cyber is not unique in experiencing market failures that prevent reasonable risk assessment and effective regulatory oversight. Indeed, there is an extensive study of this in the used-car market under the "lemon law doctrine" of economics.[8] Information asymmetry in the cyber-market means users and regulators cannot effectively identify, reward, or punish better or worse products or services. The result is that "good" cyber products may only sell for the same price as "bad" products because the usual market price discrimination incentives are ineffective. Correspondingly, information asymmetry in the cybersecurity market leaves little competitive pressure for security, and good security products sell for the same price as bad security products.

Absent effective liability protection, which is usually precluded by license disclaimers, information asymmetry effectively bars a principled determination of end-customer risks. Investment in cybersecurity by technology and service providers suffers when security benefits, which are obscured and difficult to measure, are scored against new revenue that can be obtained by equivalent investment in new capabilities. As a result, end-customers have no effective way to value better security, and providers are both unable to compete, pre-sale, on the basis of the security they offer, and they are at risk, post-sale, from unquantifiable losses because of liability protections. This situation is further complicated by the presence of adaptive adversaries whose own research investments in new attacks can serve to devalue past investments in defense.

Attacks on most engineered systems in the past were mounted on targets that were specific, well identified, and were motivated by comprehensible, predictable risk–reward assessments by attackers. They also required proximity, physical access, and specific tooling, so the risk of detection and punishment was significant. Many such attacks

---

[5] Committee on Oversight and Government Reform, 2016, "The OPM Data Breach: How the Government Jeopardized Our National Security for More Than a Generation," https://oversight.house.gov/report/opm-data-breach-government-jeopardized-national-security-generation.

[6] N. Narea, 2025, "Elon Musk's Secretive Government IT Takeover, Explained," *Vox*, February 5, https://www.vox.com/politics/398366/musk-doge-treasury-sba-opm-budget.

[7] Insikt Group, 2025, "RedMike (Salt Typhoon) Exploits Vulnerable Cisco Devices of Global Telecommunications Providers," *Recorded Future*, February 13, https://go.recordedfuture.com/hubfs/reports/cta-cn-2025-0213.pdf.

[8] G.A. Akerlof, 1970, "The Market for 'Lemons': Quality Uncertainty and the Market Mechanism," *The Quarterly Journal of Economics* 84(3):488–500, https://doi.org/10.2307/1879431.

could be thwarted by simple physical protection of high-value targets. None of these circumstantial mitigations are commonly helpful in preventing cyberattacks, especially with the proliferation of systems that are reachable through radio signals, or through vulnerable WiFi access points belonging to a remotely vulnerable close-by entity (sometimes described as a "near-neighbor attack").[9]

Attackers are often shielded from scrutiny because of the expense of investigations and, sometimes unachievable, standards of proof preventing or limiting attribution or sanction. Moreover, the existence of active attacks over a network by entities (including nation states) throughout the world makes deterrence based on legal mechanisms largely ineffective. The attackers can also plant "false flags" via use of certain tools and techniques that point investigators to a different known threat actor to avoid attribution.

A further complication is the diversity of attack surfaces—the infection vectors through which attackers and their tools engage a system. Modern AI systems, for example, can be attacked during training, while in operation, and in their delivery for use as a component in larger AI-based systems. In addition to diversity, there is also the issue of scalability of attacks. Consider, for example, the possibility of a software update that includes an adverse payload but is correctly signed by the vendor due to the theft or abuse of the code signing certificate. The scale of delivery could, for example, affect an entire fleet of vehicles or embedded CPS devices used across an entire sector.

These difficulties play a starring role in all the identified hard problems.

---

[9] S. Koessel, S. Adair, and T. Lancaster, 2024, "The Nearest Neighbor Attack: How a Russian APT Weaponized Nearby Wi-Fi Networks for Covert Access," *Volexity*, November 22, https://www.volexity.com/blog/2024/11/22/the-nearest-neighbor-attack-how-a-russian-apt-weaponized-nearby-wi-fi-networks-for-covert-access.

3

Cyber Hard Problems

Chapter 2 explored key factors underpinning difficult cyber challenges. This chapter goes on to list and discuss the list of cyber hard problems—well-defined problems where progress toward their solution would significantly improve the safety and resiliency of cyber and cyber-enabled systems, presented from the perspective of adopters and users.

CYBER HARD PROBLEM 1: RISK ASSESSMENT AND TRUST

- **The problem:** It is a huge challenge to reliably evaluate the security attributes of a proffered cyber system. There are few predictive security metrics that are reliable, quantifiable, and repeatable. This stands in contrast with simple physical systems—the tensile strength of a rope, for example, can be predicted through modeling and measured by direct testing. Making matters worse, there are readily measured system metrics, such as performance and availability, whose (relatively) simple quantifiable nature can distract from security-related metrics that are difficult to measure. Risk assessments generally need to rely on analysis of components of systems, the architecture according to which they interact, and the choices made regarding the role of the system in operational workflows—all of which influence the extent of system attack surface. An additional consideration is the nature of the threat environment associated with the system and its role; underestimating the motivation and capability of threat actors can have disastrous consequences. Information to inform these analyses can be challenging to obtain. And even when it is available, the capacity to perform this analysis in practice, even on relatively small systems, often requires enormous expertise and huge investment. Complicating matters is the non-stochastic nature of cyber risk, unlike more conventional data-informed actuarial risks. Even worse, cyber-risk assessments may feature hidden correlations (such as common components and services deep in supply chains) and secret knowledge (such as threat models and capabilities).[1]

- **Why it matters:** A tiny vulnerability in a seemingly unimportant subsystem can fully compromise the security of a poorly architected cyber system. When the overall system is used in operational contexts where outcomes are consequential and threats are significant, assurances regarding risk are difficult to provide. Risk considerations may therefore drive organizational decisions to limit the scope of application of a system within operational workflows as a means to reduce the potential consequences of attacks. This means missed opportunities to automate functions, improve efficiencies, and create new and significant cyber-enabled capabilities. The use of artificial intelligence (AI) within systems compounds this challenge (as elaborated below).

- **What makes it hard:** Traditional "black box" testing—testing without being able to see the internal elements of the subsystem—cannot ensure that correct and complete functionality of cyber systems is tested and reduces recognition of cyber vulnerabilities. Extrapolation from even the most carefully chosen test cases is unreliable due to the potential persistent state, nondeterminism in implementations, and hidden logic choices—all hidden within the box. Vendors may resist offering sufficient transparency to evaluators, which means customers and users have to make trust decisions based on judgments made either by third parties or on vendor self-attestation. The norm in vendor licenses is to disclaim liability or warranty of performance, shifting risks upward in the supply chain, ultimately to the end user, who may be in no position to make useful risk assessments—and indeed may be prohibited from doing so due to license terms. Careful auditing and analysis of audit trails can provide some basis for assessment, but systems are seldom equipped with adequate audit logging and effective analysis.

- **Where things stand today:** Users of cyber systems have to rely on a vendor's stated commitment to secure development and assessment, perhaps

---

[1] T. Tucker, 2025, "A FAIR Framework for Effective Cyber Risk Management," FAIR Institute, January 10, https://www.fairinstitute.org/blog/integrating-fair-models-a-unified-framework-for-cyber-risk-management.

augmented by red team assessments that are based on attempts to compromise the system. However, red teaming provides limited coverage, requires expertise, and is expensive. For many system elements, open-source code is often adopted because of transparency, and because users can participate directly in improving the elements. For example, web-based services are based on open-source libraries, such as NPM (node package manager), with millions of available components, making it practically impossible to assess security risks. Indeed, there is now a growing market tension, motivated by open-source issues as well as the recent cyber-focused executive orders (EOs) (such as EO 14028,[2] advancing a software bill of materials [SBOM] and zero trust) driving incrementally increasing transparency in vendor systems.

- **What is needed:** Reliable evidence-based assessment by vendors and third parties is needed to drive secure design and facilitate security evaluation. This can be enabled by incentives, such as those proposed in a March 2023 White House Cybersecurity Strategy,[3] along with evidence-based external review by third parties and, as noted, graduated degrees of transparency to acceptance evaluation focused on critical security attributes. These incentives can help drive the development of improved capabilities to create and evaluate technical evidence in support of security assurance judgments.

CYBER HARD PROBLEM 2: SECURE DEVELOPMENT

- **The problem:** Few practices exist today that can reliably ensure that software and firmware system components meet specifications of intended behaviors with respect to security-related quality attributes. Compounding this is the difficulty of developing, in the first place, policy and implementation specifications that relate to cybersecurity and resilience. As a consequence, evaluations focused on security attributes tend to be based on only fragments of direct evidence and, more extensively, on process compliance and other proxies for direct evaluation. These proxy assurances are insufficient in the face of sophisticated attackers. Compounding the problem is the cost and risk of evolving and enhancing systems, which in present practices often requires repeating a full evaluation process.

---

[2] U.S. Executive Office, 2021, "Improving the Nation's Cybersecurity," Executive Order 14028, https://www.federalregister.gov/documents/2021/05/17/2021-10460/improving-the-nations-cybersecurity.

[3] Office of the National Cyber Director (ONCD), 2023, "The National Cybersecurity Strategy," March 2, https://bidenwhitehouse.archives.gov/oncd/national-cybersecurity-strategy.

- **Why it matters:** When secure development is costly and difficult, system developers are forced to accept limits on capability and complexity for systems that are required to be deployed in consequential circumstances—where there are high standards for safety, security, and adherence to operational rules. Conversely, improvements to secure development practices can open doors—not just to more capable consequential systems but also to more rapid and affordable evolution and improvement of those systems.
- **What makes it hard:** Improvements for secure development (and evolution) range from tooling and practices to modeling and analysis techniques, language improvements, evidence management techniques, and other technical enablers, as well as skills development for developers and evaluators. The drive for capability and shorter development times—coupled with a perception that secure development is costly, time consuming, and with mostly unmeasurable results—creates counterincentives for development organizations to use secure development techniques. Developing a confidently secured component can entail deep design skills, comprehensive technical knowledge of models and analyses, and attention to details at many levels of design. It also requires, in today's practice, a base of tools, talent, and practices at a higher level than is typical. This is complicated by challenges of scalability and composability, which require close attention to technical architecture and rules of the road for application programming interfaces and frameworks (see Cyber Hard Problem 3). When hardware and firmware are involved, the challenge is amplified, especially since modern processors create new opportunities for side-channel attacks, among other risks.
- **Where things stand today:** The lack of effective measures of the various dimensions of quality and security continues to impede the creation of incentives for secure development, and the lack of incentives impedes, in turn, the creation of tools, techniques, and practices for secure development. In organizations where internal measures have been developed, there is broader adoption and advancement of practices, and with good results.[4,5] Some tools to model, analyze, and assure security properties—such as memory safety and safe control flow, for example—are now built into modern programming language designs where these properties "come for free" as language features. With these languages, developers no longer need to use separate tools or

---

[4] Amazon Web Services Cloud Security, "Provable Security Resources," https://aws.amazon.com/security/provable-security/resources, accessed February 6, 2025.

[5] S. Flur, A. Reid, S. de Haas, B. Laurie, L. Church, and M. Johnson, 2020, "Towards Making Formal Methods Normal: Meeting Developers Where They Are," *HATRA 2020: Human Aspects of Types and Reasoning Assistants*, https://research.google/pubs/towards-making-formal-methods-normal-meeting-developers-where-they-are.

even to understand the benefits and nuances of the security-relevant property. These properties include type safety (Java, Ada, and more recently Rust, TypeScript, and Hack) and memory safety[6] (e.g., Rust, Go, and Swift), and an absence of data races (Rust and, as an option, Go[7]). Adoption is extensive because the safer tools enable developers to be more productive, with error warnings earlier in development and, importantly, the new languages offer a near-identical experience to baseline tools. With regard to formal methods more generally, there is now adoption in industrial contexts where, for very specific applications, barriers of scale, usability, and affordability have been successfully overcome. Testing tools, such as fuzz testing, is now widely used because it can offer immediate benefits without wasting developer time on false positives. Chaos testing (roughly, fuzzing at architectural level) is also valuable,[8] but it requires more elaborate setup.

- **What is needed:** To have secure development, new engineering practices are needed where engineers can specify and analyze cyber components in a form that provides a measurable high level of confidence in security-relevant quality attributes. When there are practices that have associated outcome metrics, they can inform incentives for producers to use secure practices. The practices can also support composition as a pathway to scale and provide artifacts and evidence that the implementation of these components meets the specification and safety criteria. Experience has shown that techniques are used by developers and evaluators when they yield immediate productivity benefits, are easy to use, and do not require extensive setup and training. Techniques are able to support continuous evolution and evaluation of systems. There are examples where consistency is explicitly managed between the executable elements of larger-scale systems and the body of associated engineering evidence—for example, models, analyses, test cases, inspection information, and operational data. Consistency management of evidence can enable engineers to rapidly and confidently evolve secure systems, since they can more readily reuse evidence—for example, with agile-style practices such as continuous integration/continuous delivery and development, security, and operations.

---

[6] Swift, "The Swift Programming Language: Memory Safety," https://docs.swift.org/swift-book/documentation/the-swift-programming-language/memorysafety, accessed February 6, 2025.

[7] Go, "Data Race Detector," https://go.dev/doc/articles/race_detector, accessed February 6, 2025.

[8] Netflix Technology Blog, 2024, "Enhancing Netflix Reliability with Service-Level Prioritized Load Shedding: Applying Quality Science Techniques at the Application Level," *Medium*, June 24, https://netflixtechblog.com/tagged/chaos-engineering.

CYBER HARD PROBLEM 3: SYSTEM COMPOSITION

- **The problem:** Secure integrated systems are assembled from components and services of varying levels of security by programmers and designers with varying levels of knowledge and understanding, with the intent to achieve an overall secure system design. The goal of "secure composition" is to enable reliance on separately made security judgments regarding particular attributes of the individual system elements (components and services) to support efficient judgments regarding the composite system. A typical modern web application may be composed of hundreds of components drawn, for example, from the NPM open-source ecosystem, which offers more than a million components. These components can themselves be complex and composed of smaller separate components, with complex interdependencies. Another example is the architectural pattern of micro-services, which enables a more modular and scalable approach to operations on shared data. Individual examples notwithstanding, there is no comprehensive science of safe composition to guide integration, nor are there generally usable tools to validate compliance with semantic rules that allow safe composition.
- **Why it matters:** Within a safe composition framework, efforts to support security evaluations can be "reused" when the evaluated components and services are combined into diverse systems (assuming composition rules are followed). Without composition rules, which is more often the case, complex systems can be infeasible to evaluate, especially when they include complex and opaque components and services.
- **What makes it hard:** Success in composition can depend on deep technical properties of both the components to be composed and the technical design rules according to which the compositions are constructed. These properties and rules may be specific to particular security attributes, so there can be significant difficulty in achieving aggregate security judgments of fitness for use. On the other hand, and very importantly, incremental progress can be made, attribute by attribute.
- **Where things stand today:** Development of safe system composition frameworks depends mainly on talented designers and developers, as well as deep knowledge of components. But many components and services are available through development environments and repositories (such as GitHub) that can assist developers in assembling and curating components and performing some level of continuous system testing. Some of the most useful composition attributes are "hidden" in the designs of programming

languages and runtime systems—for example, "type safe" programming languages enable components to be combined by a linker into overall systems that are type safe as a consequence of the type safety of the constituent components.

- **What is needed:** Principles and architectures for system composition, advancement of expert practice in integration design, and more effective tools to support integration activities. A key enabler of composition and scale is architectural design that minimizes interactions among system elements and that enables resilience—graceful degradation in the event of compromise of a system component. That is, functions are partitioned so that an error in one partition does not adversely affect other partitions. This becomes increasingly significant with the growing scale and interconnections of systems and supply chains. Importantly, the capacity to effectively organize systems for resilience has to be planned early in development and is difficult in practice. It is also difficult to assess likely resilience outcomes early in the design process when architectural decisions are made. Chaos testing addresses this, but it occurs late in the process after systems can be tested. Resilience and composition are important areas for research focus, since "bolting on" resilience late in an engineering process is, arguably, even more challenging than bolting on other aspects of security.

CYBER HARD PROBLEM 4: SUPPLY CHAIN

- **The problem:** System elements for complex integrated systems can be sourced by a diverse array of suppliers, with components and libraries sourced from vendors, open-source projects, and custom software developers, and services sourced from diverse cloud and software-as-a-service offerings. Within subordinate supply chains, these suppliers may operate under different government or industry rules. It can be hugely challenging for an integrating system designer to readily and safely leverage the diversity of components and services required for system engineering projects, whether they are bespoke national security systems or simple commercial web applications. Supply-chain challenges are compounded by the technical challenges of composition and architecture-derived resilience.
- **Why it matters:** The diversely sourced system elements all interact within the architectural framing of the integrated system. Even when they are not opaque to analysis, typical for commercial components and services, the

interactions are difficult to model and predict. This can stymie confident acceptance evaluation, and in many cases limit the range of operational contexts within which the integrated system can be safely operated. Managing these interactions is analogous to achieving a kind of zero trust at every layer of system design, from hardware level up to major subsystems.

- **What makes it hard:** Commercial suppliers, in the interests of protecting trade secrets, generally do not want to reveal details of their components, either to end customers or to other participants in the supply chain for an integrated system. Complicating matters is that most systems employ large numbers of diverse system elements from a vast array of suppliers, commercial and open source, including both components and services. Without some degree of transparency, however, it is not possible to make sound assurance judgments at any stage in a supply chain. In many cases, transparency is provided but limited to trusted third parties or intermediaries. Even in this case, components are rarely specified in the detail required for secure composition and may only occasionally be updated or improved to address vulnerabilities as they become known. When updates are made, however, often extensive testing is required to ensure that repairs and enhancements are fully compatible with existing system elements. In the absence of updates, however, insecure components may need to be encapsulated in ways that protect other parts of the system should those components be compromised.

- **Where things stand today:** There is a movement to require end products to include an SBOM and, eventually, a hardware bill of materials. EO 14028 enshrines this intent for government systems.[9] However, SBOM data give only a hint of potential security issues in an integrated system. Open-source software provides a wide array of components that can, in principle, be directly evaluated and whose cost to maintain, evaluate, and improve is shared, but relatively few open-source projects enjoy a sufficient level of attention to code changes to assure continuing safety.

- **What is needed:** A reliable supply chain is a mechanism, supported by architectural commitments, to assure that integrated systems developers can be confident regarding diversely sourced components and services. One step along this path is for architectural decisions—at every level of design—to isolate and minimize privilege for system elements that cannot be readily evaluated or vouched for. Another step along this path is to select and curate critical open-source components that can meet the secure development desiderata. This may include investing in augmenting evidence creation in

---

[9] U.S. Executive Office of the President, 2021, "Improving the Nation's Cybersecurity," Executive Order 14028.

the open-source project to support more confident and efficient security judgments. Some industry segments, such as the automotive industry, have started to define criteria for hardware and software components as a step along this path.

CYBER HARD PROBLEM 5: POLICY ESTABLISHING APPROPRIATE ECONOMIC INCENTIVES

- **The problem:** The suppliers of cyber systems are seldom held liable even for the shoddiest products. Nor are there sufficient rewards for high quality. The well-known lemon law of economics explains why opaque cyber systems often manifest poor security as the norm. Occasionally, "brand" and informed customer demand can drive desired behavior in manifesting quality. Suppliers have avoided effective regulation with the argument that increased accountability would hinder innovation and national competitive advantage—and with an additional argument that security attributes are hard to measure, even for original developers, and therefore hard to vouch for. Many policy initiatives lead to process-focused mandates with self-attestation of compliance. These are weak proxies for delivered security in products and services. As a result, incentives have been and continue to be misaligned, leading to a market failure regarding accountable security.
- **Why it matters:** Cyber systems remain insecure, despite considerable research and attention to security practices. Consequences are experienced almost universally, from individual consumers and firms to national security applications and civil infrastructure. Even an explicitly incremental approach, taking small steps, could make a significant difference in security outcomes.
- **What makes it hard:** The complexity of cyber systems makes it technically difficult, and often infeasible, for even the best-intentioned supplier (or insurer) to warrant a system as free from important categories of defects or vulnerabilities. There is a "catch 22" situation where the inadequacy of security practices and tools impairs potential for warranting results—but without this potential, there is less incentive to invest in advancing those practices and tools. This is a measurement paradox, where the inability to directly assess levels of security in products and services can impair investment in practices to assure security—and also in ways to better assess security.
- **Where things stand today:** There are proposals, including in the 2023 White House National Cybersecurity Strategy, to create liability exposures for

vendors for certain security defects under tort law.[10,11] There are significant technical challenges. For example, the rapid evolution of systems and engineering practices would not align well with statically defined bars for liability. Additionally, the expense of litigation can often be an insurmountable barrier to be a useful remedy for most users. The bar for activating tort liability would need to be low if there are to be clear rules favoring efficient litigation. Vendors have encouraged the development of self-attested compliance to process standards as a substitute, but such compliance is not always an effective predictor of security outcomes and can be gamed by sophisticated suppliers. Additionally, compliance standards themselves can be a barrier to innovation and competition, since they focus on process rather than outcomes.

- **What is needed:** Economic incentives and transparency that can facilitate measurement and thereby enable reward for actual security, with consequences for poor security. From a legal perspective, this could include a concept of reasonable care in development based on evolving improvements in technology and, in the case of suppliers of long-lived and scalable systems, continuous improvement. The trouble is that it is difficult to provide an adequate and effective commercial legal definition of "reasonable care." Societal and commercial issues are diverse and contentious, which complicates the development of solutions.

CYBER HARD PROBLEM 6: HUMAN–SYSTEM INTERACTIONS

- **The problem:** The human interfaces presented by cyber systems are often confusing and uninformative, even to expert operators and users, and can even encourage unsafe behavior because they are poorly suited to purpose. This includes many kinds of security-related interactions, ranging from authentication and privacy control to configuration of access policies and responding to security alerts. Inadequate attention to user design for security and privacy (coupled with a failure of market incentives and regulation) has led to growing problems establishing or maintaining user security and privacy. Compounding this is the complexity and nuance of many security-related interactions between systems and humans. Indeed, errors in human interaction are by far the dominant cause of security breaches.

---

[10] U.S. Government Accountability Office, 2023, "Cybersecurity: Launching and Implementing the National Cybersecurity Strategy," GAO-23-106826, https://www.gao.gov/products/gao-23-106826.
[11] ONCD, 2023, "The National Cybersecurity Strategy."

- **Why it matters:** System engineers may have optimistic attitudes regarding the potential for training and awareness of human operators and users as a principal means to mitigate risks posed when attackers attempt to exploit human weaknesses and vulnerabilities.[12,13] The ongoing success of phishing and social engineering attacks illustrates the lack of success of this approach. In many circumstances, the mitigations most likely to make a difference are design adaptations to the human interface, including judicious choices regarding operational workflow structure and interaction design.

- **What makes it hard:** The committee sees, on the one hand, broad benefit for users to oversee decisions affecting the security of their activities and data. But, on the other hand, it can be challenging in system design to afford human operators and users the ability to make those decisions in an informed and efficient manner. In the specific case of privacy policies,[14] for example, a study found that it would take an average user 30 full working days were they to read the policies for the sites they visit over the course of a year, and presumably even more time to understand how these policies interact with each other, and the useability and usefulness situation has not improved more recently.[15] For web-browser plug-ins or other downloaded programs, there is less information available, and poor choices can lead to intrusions such as ransomware attacks. From a technology perspective, long-understood security principles, such as least privilege and auditability, are often not sufficiently respected in frameworks for browser extensions and mobile apps.

- **Where things stand today:** User studies have had a significant role in informing human-interaction design, such as in 2017 when the National Institute of Standards and Technology offered a dramatic turnabout on password guidance,[16,17] shifting from system administrator folklore regarding password construction rules to science-informed guidance. In this and other

---

[12] G. Ho, A. Mirian, E. Luo, K. Tong, E. Lee, L. Liu, C.A. Longhurst, C. Dameff, S. Savage, and G.M. Voelker, 2025, "Understanding the Efficacy of Phishing Training in Practice," *2025 IEEE Symposium on Security and Privacy* 2025:76, https://www.computer.org/csdl/proceedings-article/sp/2025/223600a076/21B7RjYyG9q.

[13] D. Lain, K. Kostiainen, and S. Čapkun, 2022, "Phishing in Organizations: Findings from a Large-Scale and Long-Term Study," *2022 IEEE Symposium on Security and Privacy* 842–859, https://www.computer.org/csdl/proceedings-article/sp/2022/131600b199/1FlQL20L5Al.

[14] S. Vedantam, 2012, "To Read All Those Web Privacy Policies, Just Take a Month Off Work," *NPR: All Tech Considered*, April 19, https://www.npr.org/sections/alltechconsidered/2012/04/19/150905465/to-read-all-those-web-privacy-policies-just-take-a-month-off-work.

[15] R. Amos, G. Acar, E. Lucherini, M. Kshirsagar, A. Narayanan, and J. Mayer, 2021, "Privacy Policies Over Time: Curation and Analysis of a Million-Document Dataset," *WWW'21*, April 19–23, https://oar.princeton.edu/bitstream/88435/pr1w562/1/PrivacyPolicies.pdf.

[16] N. Statt, 2017, "Best Practices for Passwords Updated After Original Author Regrets His Advice: Fourteen Years Later, Bill Burr Says His Tips Were Misguided," *The Verge*, August 7, https://www.theverge.com/2017/8/7/16107966/password-tips-bill-burr-regrets-advice-nits-cybersecurity.

[17] B. Fulmer, M. Walters, and B. Arnold, 2019, "NIST's New Password Rule Book: Updated Guidelines Offer Benefits and Risk," ISACA, January 1, https://www.isaca.org/resources/isaca-journal/issues/2019/volume-1/nists-new-password-rule-book-updated-guidelines-offer-benefits-and-risk.

aspects of human interaction, system engineers are starting to understand that there is not an immutable trade-off between security and usability, and that attention to good human-engineering practices, informed by empirical science, can lead to systems that are both more secure and usable.

- **What is needed:** The science of user-focused secure design needs to go beyond the focus on the user's primary functional task to encompass and enable user security with varied applications and supporting policies for data governance to enable better privacy protection.[18,19] This needs to include developing an informed understanding of the limits of human capacity to consistently adapt behavior through guidance and training. Beyond these limits, the only means to enhance security is to adapt the engineering of systems and associated workflows. An important additional element of this design challenge is the development of means by which users and operators can more readily monitor, understand, and regulate information captured and shared on their systems. A particular challenge is how to achieve this without producing information overload, introducing inefficiency, and, through this transparency, creating new avenues for attack.

CYBER HARD PROBLEM 7: INFORMATION PROVENANCE, SOCIAL MEDIA, AND DISINFORMATION

- **The problem:** Social media platforms are complex cyber systems that are used by millions of people and that can profoundly affect opinions across broad populations. These platforms provide tools that not only create social connection but also enable third parties to achieve precision targeting at scale to manipulate not only opinions but also perpetrate scams. These actions can be achieved through designed features of the business model and through use of data and deceptions to spoof content policies and controls, including deep fakes. Both aspects of this problem pose complex technical and, as widely reported, policy challenges. An additional challenge is user privacy, since online media gather highly granular usage information to inform the algorithms that tailor content to user interests and that direct advertising content.

---

[18] Conferences such as CHI and SOUPS address these issues. See Symposium on Usable Privacy and Security, 2024, "Twentieth Symposium on Usable Privacy and Security," *USENIX Security '24*, August 11–13, https://www.usenix.org/conference/soups2024.

[19] Association of Computing Machinery, 2025, "CHI Conference on Human Factors in Computing Systems," https://chi2025.acm.org.

- **Why it matters:** With proliferating deep fakes, the potential trustworthiness and reliability of online media are increasingly threatened. Without focused effort, the war of attrition between production and detection of fakes could fail. Additionally, the data gathering done by social media platforms to inform algorithms (and AI model training) can be highly granular and revelatory, with obvious privacy concerns, as evident in policy discussions surrounding TikTok.

- **What makes it hard:** The capacity for modern AI to produce deep fakes is rapidly advancing, as is the capacity to detect fakes. It is unclear whether detection capabilities can keep up. Adding to the asymmetry is the tempo of operations, which, in online media, can be very fast. Misinformation can rapidly proliferate and be precisely targeted, at which point it becomes very difficult to counter it. Also advancing, however, is the ability to watermark legitimate content such as images and videos. There are some indications that social media users can be assisted to be alert to misinformation, but this may not keep pace with the capability of producers. On the policy side, attempts to hold platform providers liable are often prevented on the basis of free speech rights and with Section 230 of the Communications Decency Act of 1996, an absence of legal incentives and, in some cases, a business model that creates counter incentives. Platforms have business incentives to maximize "engagement," which in the absence of safeguards can occur even when algorithms amplify and reinforce adverse messaging.

- **Where things stand today:** There are also potential means for image and video producers to inhibit deep-fake production using watermarks and other techniques. There is little legal liability, especially in the United States, for knowingly disseminating false information on social media. Data privacy has more policy safeguards, particularly in the European Community with the General Data Protection Regulation, but the technical security challenges remain.

- **What is needed:** On the policy side, disincentives could be enhanced for knowingly disseminating false information, and accountability could be increased regarding protection of personal data gathered in support of algorithms and advertising placement. Techniques for detecting deep fakes and watermarking original content require continual enhancement in the face of the rapid progress in generating and disseminating deep fakes, including imagery, video, voice, and other modalities.

CYBER HARD PROBLEM 8: CYBER-PHYSICAL SYSTEMS AND OPERATIONAL TECHNOLOGY

- **The problem:** Cyber-physical systems (CPS) include both operational technology—used, for example, in manufacturing, civil infrastructure, and transportation—and Internet of Things technology used in telecommunications, network hubs, security cameras, smart locks, thermostats, televisions, home, and industrial controls. CPS encompasses computing software and hardware, as well as sensors and actuators that interact with the physical world. CPS are both vulnerable and consequential. They are vulnerable because, historically, they had generally been presumed to be "off the net," and so inherited an engineering tradition with a lower standard of security than information technology (IT) systems. Additionally, CPS often lack access pathways for updates, and, finally, they can offer avenues of attack via audio, network, and physical access.

- **Why it matters:** CPS are highly consequential because they are the control fabric for civil infrastructure systems, industrial controls, and manufacturing systems, as well as embedded national security systems. They are also now critical infrastructure in homes and offices. There have been many significant attacks over the years in nearly all of these sectors. That they are largely invisible, as if part of the furniture, often causes them to be ignored.

- **What makes it hard:** Secure design of a CPS requires both hardware and software expertise. CPS deployments generally do not have access to trained professionals prepared to work on these systems, and many do not support remote update should repairs need to be made or vulnerabilities patched. This means that the technical challenges of reengineering hardware and software for security are compounded by the high stakes of "getting it right" at the outset. Additionally, many of these systems feature real-time controls, which can create a tangle of internal interdependencies that complicates reengineering for security and resilience, both for individual CPS devices and at scale for distributed networks.

- **Where things stand today:** There is technical attention to CPS security, but challenges remain to develop real-time systems that are more modular, scalable, and secure.

- **What is needed:** CPS are important targets, and consequently require all the protections of modern IT systems—as well as the requirements previously noted regarding transparency, configuration integrity, critical evaluation, secure design and composition, and secure supply-chain requirements. Specific focus on CPS is important because many of them provide real-time control, which leads to both design and assurance challenges. Engineering practices are also an issue, because many embedded systems need support for rapid remediation of issues, including critical infrastructure and weapons systems.

CYBER HARD PROBLEM 9: ARTIFICIAL INTELLIGENCE AND EMERGING CAPABILITIES

- **The problem:** Contemporary AI is at the leading edge of a technology storm for which it may be difficult to predict the full set of safety and security ramifications. Today, AI is generally based on neural-network models supporting machine learning (ML) and generative AI using statistical inference based on large corpora of training data. Despite widespread adoption, protection criteria for modern AI technologies and systems are still nascent.[20] Efficacy is dependent on many factors including the reliability of training data, but these data are often opportunistically acquired and poorly curated, biased, and inaccurate. Dependence on training data introduces complex issues beyond quality and security, including copyright and data ownership. Another factor is that even sophisticated users may attribute characteristics to modern AI systems based on sampled experience, which may not reveal the diverse kinds of weaknesses and vulnerabilities that are often present. This may become problematic from both a security and safety perspective as traditionally human-centric tasks are automated by the use of AI. (That AI can be used to automate offensive cyberoperations is addressed separately in the section "Defense Against Offensive Artificial Intelligence" in Chapter 4.)
- **Why it matters:** Modern AI models are becoming significant elements of a growing range of systems. AI has a wide range of current applications and an enormous range of potential applications. Although these include some safety critical applications such as transportation and medical image analysis, the AI systems are generally in advisory roles where accountability resides with human operators. But there are aspirations to apply modern AI in contexts where trustworthiness is essential, including faster-than-thought autonomy, technical systems engineering, and a range of expert applications.
- **What makes it hard:** Security depends on a carefully developed, principled framework where weaknesses and vulnerabilities can be identified

---

[20] National Institute of Standards and Technology, 2024, "AI Risk Management Framework," https://www.nist.gov/itl/ai-risk-management-framework.

and clearly tied to potential adverse outcomes. In ML, a vast corpus of training data, often undifferentiated with respect to quality and correctness, are transformed into neural logic in the form of up to hundreds of billions of parameters in a neural network. This creates an opacity to analysis such that, even with full transparency of the system and its parameters, the behaviors of neural networks cannot be reliably predicted. Even with carefully curated training data, the statistical nature of neural-network models means that outputs, whether from ML models or generative AI models, can be inexact and untrustworthy. A statement "the neural network will never do this" is often impossible to assure. Thus, the human factor compounds the difficulty of the problem due to automation bias—the propensity for humans to favor decisions from autonomous systems. The security of the applications also resurfaces old security challenges in this new paradigm. For example, applications leveraging generative AI may be susceptible to SQL-like injections from untrusted sources (indirect prompt injection) because they leverage the same input pathways for a neural network for both "data" (untrusted documents) and "code" (textual instructions), violating a core security principle. Making matters worse is the extraordinary breadth of attack surface, encompassing training data, network architecture, and operational inputs—all in addition to traditional cybersecurity attack surfaces associated with AI models as software components and services.

- **Where things stand today:** In the familiar "opportunity and challenge" framing, modern AI systems are extreme in both their great promise and great potential peril. Current generative AI systems can provide both amazing insights, due to their extraordinary powers of recall over the vast ocean of their training data and information sources provided at runtime, and also laughably incorrect hallucinatory conclusions, due to the fundamentally approximate representation of the "knowledge" embodied within a network (including falsehoods, irony, and AI outputs), and the stochastic nature of response generation. There are, however, a wide range of advisory applications where modern AI systems work very well. Adaptations to both the systems (e.g., plug-ins, retrieval augmented generation, agentic systems, and similar techniques) and the operational workflows that incorporate them (e.g., protocols for human operators and guidance for human users) enable systems developers to exploit the strengths of these models. It is, however, the early days for AI, and there are no agreed-upon security foundations.
- **What is needed:** As a rapidly evolving automation technology that is powerful but not easily predictable, software developers that leverage AI must place

explicit human accountability and safe human outcomes as primary design priorities. There are many steps that can be taken with modern neural-network models to make incremental improvements, but it is important to note that many of the weaknesses and vulnerabilities are intrinsic to the neural-network technical architecture. As such, foundational improvements in AI modeling or model augmentation are needed to make models more auditable for reliable use as system components. Paths forward may include the hybrid use of neural networks with symbolic techniques, wherein models are augmented with explicit knowledge graphs and logic-based deductive processes. Since at present, models themselves are not auditable, it is important, not just to identify weaknesses and vulnerabilities in modern AI, but to develop a principled security practice that includes modeling and analysis techniques for detection and mitigation. Applying secure design principles (e.g., access controls, trust boundaries, and data/control separation) may remediate many security and safety challenges with generative AI applications, but appropriate incentives are apparently still required and frameworks for developing generative AI systems that explicitly promote or enforce them. Lastly, as general-purpose knowledge systems that often include generic built-in safety and security guardrails, improvement to application-specific safety and security guardrails is required. This should include supplementary external guardrails independent from those trained explicitly into a model so that controls can be scoped and enforced independently and provide capability parity to the generative model.[21]

CYBER HARD PROBLEM 10: OPERATIONAL SECURITY

- **The problem:** There is increasingly pervasive dependence on large-scale systems such as cloud infrastructure from Amazon Web Services, Azure, and Google, as well as scaled applications such as search and email from Microsoft and Google. For these large-scale systems, users have little insight, beyond vendor self-attestation, into the state of security configuration. These examples are of mainstream consumer applications, but similar considerations arise within the IT infrastructure of large organizations including, for example, enterprise resource planning systems.

---

[21] A. Wei, N. Haghtalab, and J. Steinhardt, 2023, "Jailbroken: How Does LLM Safety Training Fail?" *37th Conference on Neural Information Processing Systems (NeurIPS 2023)*, https://papers.nips.cc/paper_files/paper/2023/file/fd6613131889a4b656206c50a8bd7790-Paper-Conference.pdf.

- **Why it matters:** As systems grow in scale, interconnection, and significance to their users, resilience becomes increasingly important, with concerns over confidentiality, integrity, and availability for both data and operations. Resilience to attack and failure can include techniques to isolate compromised system elements to enable the system to continue to operate through the compromise, analyze the situation, and subsequently recover. Defined processes for operational situational awareness and remediation are increasingly critical for practical operational security as well as assessment, both real time and forensic. This is not possible unless attacks can be detected quickly, potential consequences assessed rapidly, and defensive actions initiated appropriately.
- **What makes it hard:** Operational security for larger organizations poses multiple challenges—prevention, detection, response, and recovery. Anticipation and mitigation mean reducing both potential for successful attacks and also the extent of consequences when attacks (and failures) result in compromises. These can involve significant planning and cost. Detection and response, for example, require threat intelligence, constant practice, the diligent development of supporting tools, and comprehensive knowledge of overall organizational networks and applications. Detecting attacks and determining consequences can be expensive and error prone. Many government systems, for example, include diverse elements from diverse vendors, including security support. Recovery from attacks is an important element of security and, as has been shown by numerous ransomware attacks, many organizations struggle to accomplish this. Additionally, many societal systems depend on a few key infrastructures, such as cloud-based services, which can become single points of catastrophic failure, with risks of compromise, availability, and integrity. High levels of operational security require organizational maturity and top-tier support due to costs, complexity, and the need for architectural control. Business and market incentives pose challenges that are amplified by the difficulty of assessing capacity with respect to the several dimensions of operational security.
- **Where things stand today:** There are large organizations that have developed well-tuned processes, continuous monitoring and assessment, as well as effective tools to manage at-scale systems (and complex configurations) that include engineered reliability. In many cases, this knowledge is closely held and difficult to assess from the outside. Operational security for smaller organizations (and individuals) remains hit or miss.
- **What is needed:** Systems need to be designed to support and anticipate operational security needs, including detection of attacks via automation and

automatic assessment of the potential consequences of attacks and remediation actions. Despite the typically unique organization-specific designs of large operational systems, there are common principles of design and operations that can be identified, applied, and measured.

Readers will easily recognize that these hard problems are not independent of one another and cannot be solved individually. Therefore, in highlighting them, the committee also hopes that collective action can be organized across government, industry, and research communities to make progress addressing them.

4

The Producer Perspective

Chapter 3 lists cyber hard problems from what might be called the point of view of a "consumer" relying on cybersecurity in a product or service. This provides a tidy taxonomy but does not describe specific issues or problems that, if addressed, would represent significant progress toward solving them. This chapter describes these specific challenges, which can also be seen as the perspective of the "producer" who needs a well-characterized set of independent principles and procedures—technical, policy, and operational—that are prerequisites to addressing the cyber hard problems.

The consumer and producer lists do not map neatly to one another for the following reasons:

- Solving a consumer cyber hard problem may require solving many different (e.g., technical) producer problems, and the technical problems may affect many different consumer-level problems.
- The consumer cyber hard problems themselves are interdependent. For example, risk assessment depends on secure design and composition as well as supply-chain integrity, the availability of metrics and the reliability of data and information and is profoundly affected by policy related to "remedies and incentives" that help making such an assessment and being able to reasonably rely on it.
- Policy and economic incentives affect nearly all the producer cyber hard problems since incentives determine what producers do (or should do), what prices they can charge to consumers, and what redress consumers have if they fail to do so.

- Security metrics (if they exist) and the related ability to determine the cybersecurity of a system at a reasonable cost, based on easily obtainable measurements or information, affect producers because it would allow them to determine whether they have succeeded in "secure, resilient design and composition" as well as determining the security characteristics of their own supply chain and risk of liability. Finally, the consumer's ability to determine whether a system was securely designed or implemented and whether its operation is secure "in practice" would best be based on metrics instead of largely unsupported assertions by producers.

Some of the new cyber hard problems, such as the integration of hardware and software into a cyber-physical system (CPS) or artificial intelligence (AI), depend on essentially all of the producer cyber hard problems. However, they bring new important subproblems that are critical and unsolved.

For clarity and brevity, the producer cyber hard problems are described below in terms of concrete functional, operational, new technology, or policy problems. Some *producer* cyber hard problems may seem to duplicate *consumer* cyber hard problems. For example, "secure, resilient design" from a consumer's point of view involves solving many subproblems because it is a characteristic of an entire cyber system and its operation. The "secure design" problem in this section addresses the technical problems involved in designing a more or less fully specified system (development tools, testing tools, design practices, needed workforce competencies, etc.). This also applies to secure composition. Accurate provisioning of data and information may be directly visible to a consumer, but it may be a characteristic of the training data used to produce an AI model, which may be of no direct interest to a consumer.

Some of the producer cyber hard problems can be solved in a rather satisfying manner by principled techniques that often go under the rubric "the science of security." Examples include complete access and information flow models.[1] This sort of solution is the "gold standard" for scientific progress but has only been applied to very carefully described and constrained subproblems.

FUNCTIONAL CYBER HARD PROBLEMS

Functional cyber hard problems deal with the design of secure, interoperable products and infrastructure.

---

[1] F.B. Schneider, 2012, "Blueprint for a Science of Cybersecurity," *The Next Wave* 19(2):47–57, https://www.cs.cornell.edu/fbs/publications/SoS.blueprint.pdf.

Cloud Visibility and Centralized Risk

The emergence and popularity of cloud computing are due to convenience, efficiency, and (in some cases) cost-saving. However, while almost everyone uses or relies on cloud computing services today, there are few dominant providers, and their operations are unknown to outsiders. The monoculture and opacity result in key challenges to building and operating resilient cloud systems.

Many of the drivers for cloud computing adoption involve cost and convenience, but the trade-off most pertinent for security is control and understanding. While models of shared responsibility usually exist between the customer and provider, they are not always consistent or complete. Multi-tenancy in a provider's environment can affect the visibility available to each customer when the provider cannot separate logging, back-ups, or forensic data for each tenant; sometimes the physical location of a provider's data center is confidential. This is the "isolation" problem in shared resources. Shared hardware increases the risk of lateral movement by an attacker from one customer to another, and the network traffic needed by one customer may require that the provider cannot block some network traffic even if some tenants want them to. Finally, depending on the service, customers may not have visibility into traffic and interactions that happen within the provider's environment, only the traffic that happens directly between the customer's instance and the customer's own location. Business email compromise (BEC) for the purpose of redirecting payments is a big problem and shows no sign of slowing.[2] It is often accomplished by adding filtering rules redirecting emails pertinent to payments to a scammer acting as a "man in the middle" who instructs the payments to be redirected. Outsourcing email to "the cloud" where such critical rule changes for a single email user may not be quickly caught is a common problem.

Besides visibility issues, another hard problem for cloud computing is orchestration. It is difficult to find reliable statistics on multi-cloud use that do not come from a single cloud provider, and control policies between providers can be inconsistent. A look at the latest Cloud Controls Matrix[3] from the Cloud Security Alliance tells the story: 197 control objectives in 17 domains. Organizations face security challenges regardless of the strategy they embrace. Using one provider can risk a single point of failure,[4] and

---

[2] P. Harr, 2024, "The Weaponization of AI: The New Breeding Ground for BEC Attacks," Forbes Technology Council, June 14, https://www.forbes.com/councils/forbestechcouncil/2024/06/14/the-weaponization-of-ai-the-new-breeding-ground-for-bec-attacks.

[3] K. Rundquist, 2024, "Cloud Security Alliance Announces Implementation Guidelines v2.0 for Cloud Controls Matrix (CCM) in Alignment with Shared Security Responsibility Model," *BusinessWire: Cloud Security Alliance (CSA)*, June 4, https://www.businesswire.com/news/home/20240604212963/en/Cloud-Security-Alliance-Announces-Implementation-Guidelines-v2.0-for-Cloud-Controls-Matrix-CCM-in-Alignment-with-Shared-Security-Responsibility-Model.

[4] Intelligent Transportation Systems Joint Program Office, "ITS Deployment Evaluation," Department of Transportation, https://www.itskrs.its.dot.gov/2019-l00856, accessed December 5, 2024.

using more than one provider incurs complexity and management costs (as well as the increased systemic risk of any one provider having an outage).

A particular risk with cloud computing is the durability of artifacts; although using only what you need, when you need it can help reduce costs, the practice requires more rigor in managing the life cycle of those instances. For example, the Colorado Department of Transportation fell victim to a ransomware attack in 2018[5] when a virtual server was not secured properly because it was intended to be temporary, and yet it was connected to the agency's active directory domain, which allowed the attacker to gain additional privileges. Retention periods for backups and logs are other examples of critical artifact management properties.

The start of the COVID-19 pandemic in 2020 forced more organizations to embrace cloud computing with remote access, which, in turn, drove the development of architecture changes such as "edge computing." Remote user traffic that had to pass through on-premises infrastructure to access cloud-based resources resulted in network bottlenecks and latency; the Secure Access Service Edge emerged in response, putting the users and resources closer to one another. All this reliance on third-party providers has opened new areas of attack as well as complicated security management;[6] because any given end-to-end interaction now involves additional personnel, terms of service, and levels of visibility and control different from a simple governance model.

Assessing how well a provider secures its offerings is one challenge covered further in the section on supply chain security below. A related cyber hard problem is the process of incident response, for the reasons outlined above. Putting these together creates an overall cyber hard problem that deserves attention because one provider's outage or compromise can affect literally thousands or even millions of customers. The notion that every organization is solely responsible for securing itself is outdated, as is acknowledged in the most recent White House strategy document cited earlier. The answer lies not just in another technology product that customers must layer on top of the already complex infrastructure (complexity to manage complexity), but includes aligned incentives and clearer, more consistent security standards and responsibilities for these "linchpin" cloud providers.

Cloud infrastructure is included in cyber hard problems 1, 2, 3, 4, 5, and 10.

Identity, Authentication, and Access Control in the Context of the Global Commons

The democratization of technology has increased the number of accounts each user relies on dramatically, but the demographics of those users have changed. Consumers as

---

[5] Ibid.
[6] Office of the National Cyber Director (ONCD), 2023, "The National Cybersecurity Strategy," The White House, March 2, https://bidenwhitehouse.archives.gov/oncd/national-cybersecurity-strategy.

young as 3 and as old as 103 must (securely!) authenticate themselves for purposes as varied as online games, banking, education, government services, medical care, and employment. Attackers have taken advantage of the gaps in implementation and differing levels of user sophistication by capturing credentials or tricking the legitimate account holder into exposing authentication information. Identity, authentication, and access control have become a significant battlefield and a favored "soft target" even as viable technical solutions have been accepted.

In response, identity and access management (IAM) technology has evolved to meet this need.[7] IAM trends include the following:

- Increasing the availability of two-factor and multi-factor authentication (2FA and MFA, respectively). Although 2FA was available long before the 2005 cyber hard problems list, more tools emerged with different ways of enforcing the "something you have" that contributes to more secure authentication. At the same time, the use of SMS as one of these authentication factors, although it is the easiest and cheapest to deploy at scale, particularly where feature phones (precursor to the "smart" phone era) are more widely available than "smart" phones, has come under widespread attack through both social engineering and the practice of "SIM swapping."[8]
- The growth of single sign-on (SSO) to evaluate and pass authentication data under time-limited conditions so that the user does not have to repeatedly authenticate to access multiple systems, applications, and data in the scope of a single task or enterprise. This generally takes place within one governance area (such as a corporate infrastructure), where a consistent access policy can be enforced; sometimes SSO can be federated among disparate cooperating entities.
- The implementation of passkeys (cryptographically generated key pairs) that are securely stored within the user's device or other hardware, taking the place of memorized (and usually overused and oversimplified) passwords.
- The development of password managers, which store and in some cases autofill passwords at authentication time, making it easier for users to choose complex or unique passwords without having to remember them or type them in.

---

[7] National Institute of Standards and Technology (NIST), 2023, *Digital Identity Guidelines*, SP 800-63, https://pages.nist.gov/800-63-3.

[8] Federal Bureau of Investigation, 2022, "Criminals Increasing SIM Swap Schemes to Steal Millions of Dollars from US Public," Public Service Announcement: Alert Number I-020822-PSA, February 8, https://www.ic3.gov/PSA/2022/PSA220208.

- Heavier use of biometrics (fingerprint, facial recognition, gait analysis, typing analysis, and other techniques) for the initial authentication in place of passwords, to unlock other stored credentials such as passkeys, or coercion detection.

Around the early 2000s, the UK-based Jericho Forum proposed a stronger authentication model, called a "collaboration-oriented architecture." With the principle of explicitly authenticating every access request, regardless of where it originated, "zero trust"[9] resulted in many additional authentication factors being developed, such as GPS-based location, just-in-time analysis of the security state of the device being used for access, WiFi fingerprinting, biometrics, passkeys, and more. The additional options in this factor portfolio also made it more difficult to build and test consistent access policies.

Another complicating factor in IAM is time. Authentication has moved beyond a one-time event into continuous evaluation of the user's factors, including location or security state changes during the session and activity alerts. Depending on the assessed risk and policy, the system might invoke a step-up authentication process with more factors to ensure that the access is legitimate. The continuous assessment may also take specific events into account, using data received either from the system owner's own infrastructure (such as network telemetry, application changes, or a change in user access from the identity provider) or from collaborating entities. For example, a password change should invalidate all currently open sessions. The OpenID Foundation Shared Signals working group[10] is tackling the challenge of standardizing and sharing access-related events, but as with many information-sharing initiatives in cybersecurity, misaligned incentives can hamper this goal.[11] Finally, there is the concept of granting authorization dynamically, in a just-in-time fashion, rather than equipping an account with static permissions. The "zero standing privileges" approach is intended to harden existing user accounts in the face of attacks but carries its own set of associated management complications.

Identities themselves are now more widespread and context-specific than they were 20 years ago—they involve not just "who are you?" but "why should you have access to this particular resource at this point in time?" Establishing the right to access by verifying that the user is a citizen, a parent, a doctor, a partner, an employee, a customer, or a student can require collecting attributes from many different trusted parties.

---

[9] S. Balaouras, J. Blankenship, D. Holmes, P. McKay, J. Burn, A. Tatro, and M. Belden, "The Business of Zero Trust Security," Forrester, https://www.forrester.com/zero-trust, accessed February 6, 2025.

[10] T. Cappalli, S. Miel, S. O'Dell, and A. Tulshibagwale, "Shared Signals Working Group—Overview," *OpenID*, https://openid.net/wg/sharedsignals, accessed February 6, 2025.

[11] Any time information sharing is voluntary, commercial drivers can get in the way of sharing useful and complete information. For example, security vendors with their own threat intelligence teams may avoid sharing unique data if it is seen as a competitive advantage or delay release past the point of timeliness in order to publish within marketing schedules.

The collection of personal data, often incentivized financially for marketing and resale, contributes to the attack surface for each individual, as threat actors can obtain a wider variety of demographic data and secrets needed to register, use, and recover access.[12]

Collecting this verified data, storing them securely, and only releasing them where necessary are all associated privacy challenges. Proposed solutions include a digital self-sovereign identity framework, such as the European eIDAS regulation[13] that will require every European Union country to offer a digital identity wallet by 2026. U.S.-based government services such as Global Entry are now offering digital IDs for mobile devices. Large-scale public identity providers such as Apple, Facebook, and Google have been offering to ease usability for consumers by letting them use their account identities for logging in to other sites, making payment transactions, and so on. However, these varied offerings come with their own governance and privacy goals, which the general public may not be able to evaluate.

To make matters even more complicated, identity management and governance have moved beyond the realm of humans. Machine identities, workload identities, and the operational system accounts that underpin all types of infrastructure, from applications to network routers, all need to be addressed in a coherent way, particularly with the growth of the Internet of Things (IoT). Wherever access is not tied to an individual human, or wherever two entities communicate with one another without human initiation, the authentication, authorization, and identity issues still apply. The Workload Identity working group[14] is addressing some of these issues, but the drudgery of tracking, auditing, and protecting dormant on-premises system accounts remains with the owners of the infrastructure. Not only do attackers regularly target default passwords on these systems, but the potential areas for attack now range from critical infrastructure (utilities, nuclear power plants, 911 systems, medical equipment) to security cameras, home thermostats, baby monitors, and indeed anything that is connected to the Internet under the guise of being "smart." Non-human identities are equally important to combat cases where an attacker simulates a website or message to trick a user into supplying credentials; the machine needs to authenticate itself to the human.

One final point is that although IAM frameworks and technology have evolved, they are also extending the "long tail" of legacy systems that are too costly to retrofit. For every web-based application that now uses passkeys for authentication, there is also a decades-old banking mainframe or industrial storage tank that must still interoperate.

---

[12] Department of Defense (DoD), 2023, "2020 DSB Summer Study on New Dimensions of Conflict: Executive Summary," DoD Office of Prepublication and Security Review 23-S-2072, April, https://dsb.cto.mil/wp-content/uploads/reports/2020s/DSB-SS2020_NewDimensionsofConflict_Executive%20Summary_cleared.pdf.

[13] European Commission, 2024, "eIDAS Regulation," April 4, https://digital-strategy.ec.europa.eu/en/policies/eidas-regulation.

[14] IETF Datatracker, "About the IETF Datatracker," https://datatracker.ietf.org/release/about, accessed February 6, 2025.

The overarching technical hard problem within IAM is configuring and managing policy. Because of complicated mechanisms for performing authentication and authorization, expanded user demographics, and many, often ephemeral, systems that require IAM without governance, the need to protect these vulnerable attack points requires simplifying and possibly centralizing policy management. Today's chief information security officer has no straightforward way to decide which factors to use, how to model the operational impact of a policy change, how to negotiate policy enforcement with external providers, or even how to get all the event data they need to make (or possibly automate) risk decisions involving policy across the entirety of the technologies and environments. One example of the current fractured state of IAM is universal access revocation, sometimes called "single logout"—the problem of identifying all accesses belonging to a departing user and revoking them, terminating any existing live sessions, and handling deeper layers of associated application and system access such as cryptographic tokens. This "holy grail" of IAM applies not only to access management in workplace scenarios but also to any active incident response involving a compromised user account. Achieving it comes with trade-offs—for example, tracking every place where a user is active can also have privacy implications.

As mentioned above, the nontechnical hard problem aspect of IAM involves governance. Identity management, authentication, and access control are driven by commercial entities and are fragmented accordingly, as governance falls to a population of private and public resource owners, not simply a central one in the role of an enterprise employer as in the past. In countries where each citizen has a single government-managed digital ID, the resulting centralization affords better technical solutions and enhances accessibility for underserved populations. Long-standing mistrust of centralized government in the United States stands in the way of creating a centralized digital identifier. Where cultural distrust of centralization is higher, it may be more practical to develop a broad federation model, allowing disparate resource owners and consumers to use a consistent and reliable framework for negotiating IAM features and processes.

Access controls affect cyber hard problems 1, 2, 3, 4, 6, 7, and 10, although it affects others to a more limited extent.

Developing an Empirical Basis for Security Decisions

Cybersecurity is the property of technological artifacts, people, and processes to resist attacks by an adversary. At every stage of a system's life cycle, including design, implementation, acquisition, testing, deployment, training, use, monitoring, maintenance, and retirement, there are a broad range of decisions to be made that will influence these properties. What authentication architecture should be chosen? What programming languages, tools, and processes should be used to minimize the introduction of implementation bugs? When acquiring such a system, how should an organization compare the security it offers to competing systems? How should their information technology (IT) professionals configure the system to support its security assumptions? How should employees be trained to use it? When must a system be updated or retired due to security liabilities? Such questions, explicit and implicit, are being answered thousands of times a day. It is widely held that some of these choices are likely better than others. Indeed, many believe that there may be "best" choices for a given situation and that certain decisions, if taken, would seriously foreclose attacks.

While it is tempting to hope that these questions might be answered a priori—that with the proper levels of formal reasoning, systems might be designed and proven secure against reasonable threat assumptions—such results are rarely available. Real systems operate in a messy world, typically more complex than can be modeled, with countless deviations from idealized abstractions, with multiple humans in the loop, and adversaries who formulate their attacks based on the assumptions made by defenders.

This is strong motivation to place cybersecurity on a firm empirical footing—akin to evidence-based medicine—where careful data collection and analysis can differentiate and prioritize among the plethora of factors and approaches. However, a perennial challenge for the cybersecurity community has been to establish a rigorous evidentiary basis for evaluating such choices in a way that predicts outcomes. As a result, most of today's established cybersecurity "best practices" are based on a combination of perceived common sense and received wisdom.

There are many reasons to question the quality of this status quo decision making.

- *Disagreement about best practices.* Even among experts, there is frequent disagreement about the set of best practices, and which are the most important. For example, Redmiles and colleagues' recent large-scale analysis of online security and privacy advice identified 374 distinct pieces of guidance, of which a set of experts surveyed identified 188 as being among the "top 5" practices that should be followed.[15] Using a similar methodology, Reeder and colleagues surveyed 200 security experts who identified 152 distinct "top 3" practices.[16] While consensus is no substitute for evidence, a lack of consensus suggests an information-poor environment in general.
- *Poor evidence even where there is consensus.* In those cases where there has been agreement about best practice, it is frequently based on simplified abstractions of how systems are built, used, and attacked. Empirical data are

---

[15] USENIX, 2020, "29th USENIX Security Symposium," August 12–14, https://www.usenix.org/conference/usenixsecurity20.

[16] IEEE Symposium on Security and Privacy, 2017, "38th IEEE Symposium on Security and Privacy," https://www.ieee-security.org/TC/SP2017.

ultimately needed to evaluate the impact of these best practices. For example, many major software developers follow some sort of secure development process (e.g., incorporating threat-modeling, code review, fuzz testing, such as Microsoft's Security Development Lifecycle), but even after 20 years of experience, little is known about how effective these processes are for improving security. A 2019 Dagstuhl Seminar summarized the situation as follows: "There is little empirical data to quantify the effects that these principles, architectures and methodologies have on the resulting systems."[17] Similarly, there is broad acceptance of the security benefits of outsourcing key services to cloud providers with high-quality security teams, but there is no practical way to reason about how to compare these benefits with the independent risk of correlated losses when one of these cloud platforms is itself compromised.

- *The evidence that is collected repeatedly contradicts prior assumptions.* In the handful of cases where there have been extensive empirical analyses of these assumptions, best practice has repeatedly been found wanting. For example, for decades the received wisdom concerning password content requirements was that requiring more character classes and longer passwords was superior, on the assumption that this would increase entropy and thus increase the work factor for an attacker to guess. Password reset policies (e.g., resetting passwords more frequently) was deemed superior, on the assumption that it reduces the window of vulnerability for using a password. While these rules are frequently ascribed to Bill Burr's 2003 authorship of National Institute of Standards and Technology (NIST) SP 800-63 and its appendix that codifies these arguments,[18] the history is significantly older and more diffuse. For example, these issues are discussed in Morris and Thompson's "Password Security: A Case History"[19] and 20 years later in Adams's and Sasse's "Users are Not the Enemy."[20] However, regardless of the origin, the key point is that these arguments have been widely accepted and implemented based on their implicit assumptions concerning attacker and user behavior, without any significant empirical scrutiny. Yet after a decade of research, it became clear that these approaches were in fact not superior in practice, had been based on incomplete assumptions about how users and attackers work, and frequently

---

[17] A. Shostack, M. Smith, S. Weber, and M.E. Zurko, 2019, "Empirical Evaluation of Secure Development Processes," *Dagstuhl Reports* 9(6)1–25, Schloss Dagstuhl – Leibniz-Zentrum für Informatik, https://doi.org/10.4230/DagRep.9.6.1.

[18] R. McMillan, 2017, "The Man Who Wrote Those Password Rules Has a New Tip: N3v$r M1^d!" *Wall Street Journal Pro-Cybersecurity*, August 7, https://www.wsj.com/articles/the-man-who-wrote-those-password-rules-has-a-new-tip-n3v-r-m1-d-1502124118.

[19] R. Morris and K. Thompson, 1979, "Password Security: A Case History," *Communications of the ACM* 22(11):594–597.

[20] A. Adams and M.A. Sasse, 1999, "Users Are Not the Enemy," *Communications of the ACM* 42(12):40–46.

interact with usability in ways that actually reduce security. A variety of recent studies have been unable to find clear and convincing empirical evidence that well-accepted practices actually improve security, including embedded phishing training,[21] indicators of compromise-based threat intelligence sharing, and prompt browser updating.[22] Some studies support actual benefits, but with only minor effect (e.g., Thompson and Wagner's study of code review impact on security vulnerabilities[23]). For the vast majority of security decisions made, there are no studies at all. We simply act without evidence.

This is far from a new realization. "Metrics for Security" was identified as a key cyber hard problem in the 1995 version of the InfoSec Research Council's Hard Problem List,[24] the 2002 National Research Council consensus study report *Cybersecurity Today and Tomorrow: Pay Now or Pay Later*,[25] the Computing Research Association's 2003 "Four Grand Challenges in Trustworthy Computing,"[26] the 2005 President's Information Technology Advisory Committee report *Cyber Security: A Crisis of Prioritization*,[27] and again in the InfoSec Research Council's 2005 re-up of Hard Problem List[28]—relabeled as "Enterprise-level Security Metrics" (although not because the smaller scale problems had been solved, indeed that study states, "Most of the existing [security] metrics are of questionable utility, even with respect to individual software systems."[29,30]). Indeed, almost 20 years later, the software-focused 2024 Office of the National Cyber Director (ONCD) report *Back to the Building Blocks: A Path Toward Secure and Measurable Software* opines that still "it is critical to develop empirical metrics that measure the *cybersecurity quality* of software."[31]

---

[21] D. Lain, T. Jost, S. Matetic, K. Kostiainen, and S. Capkun, 2024, "Content, Nudges and Incentives: A Study on the Effectiveness and Perception of Embedded Phishing Training," arXiv:2409.01378.

[22] L.F. DeKoven, A. Randall, A. Mirian, G. Akiwate, A. Blume, L.K. Saul, A. Schulman, G.M. Voelker, and S. Savage, 2022, "Measuring Safety Practices," *Communications of the ACM* 65(9):93–102.

[23] C. Thompson and D. Wagner, 2017, "A Large-Scale Study of Modern Code Review and Security in Open Source Projects," *PROMISE '17*, November 8, https://people.eecs.berkeley.edu/~daw/papers/coderev-promise17.pdf.

[24] The 1995 Infosec Research Council (IRC) *Hard Problems* is not easily found, but the problems themselves are available in Appendix A, "Retrospective on the Original Hard Problem List," of the 2005 *Hard Problem List* report. See IRC, 2005, *Hard Problem List*, November, https://www.nitrd.gov/documents/cybersecurity/documents/IRC_Hard_Problem_List.pdf.

[25] National Research Council, 2002, *Cybersecurity Today and Tomorrow: Pay Now or Pay Later*, National Academy Press, https://doi.org/10.17226/10274.

[26] Computing Research Association, 2003, "Four Grand Challenges in Trustworthy Computing," https://archive.cra.org/Activities/grand.challenges/security/grayslides.pdf.

[27] President's Information Technology Advisory Committee, 2005, *Cyber Security: A Crisis of Prioritization*, National Coordination Office for Information Technology Research and Development, February, https://www.nitrd.gov/pubs/pitac/pitac_report_cybersecurity_2005.pdf.

[28] IRC, 2005, *Hard Problem List*.

[29] IRC, 2005, *Hard Problem List*, p. 56.

[30] D. Maughan, 2006, "Infosec Research Council Hard Problem Lists," Department of Homeland Security, Science and Technology Directorate, January 26.

[31] ONCD, 2024, *Back to the Building Blocks: A Path Toward Secure and Measurable Software*, The White House, February, https://bidenwhitehouse.archives.gov/wp-content/uploads/2024/02/Final-ONCD-Technical-Report.pdf, p. 11.

Given the clear need, why, then, has there been such limited progress in cybersecurity while other fields, such as medicine, have been able to incorporate empirical data to great success?

The traditional refrain, well described in Herley and van Ooorschot's overview paper on the "science of security,"[32] is a triumvirate of problems that describe the "unique challenge" in measuring cybersecurity.

- *Adaptive intelligent adversaries.* This concern is, by far, the most common—the fact that the scope of adversarial behavior cannot be empirically accounted for in measurements of systems. An adaptive adversary, by definition, can change their method of attack—either in response to changes in the system itself, or changes in attacker motivation, knowledge, or investment. Thus, evidence concerning the cybersecurity of a system at a point in time may fail to predict changed outcomes driven by an attacker's new behavior or abilities.
- *Lack of fundamental laws.* Unlike the invariant natural laws of physics, cybersecurity is a moving target—"too entwined with human behavior and engineered systems to have universal laws."[33] Thus, it is believed that few aspects of system behavior extracted from empirical observation can be effectively generalized into firm rules that may be counted on for future decision making.
- *Dynamic artifacts.* The modern computing environment is constantly changing, as are the behaviors and modes of use of its users. Thus, at best, one can obtain evidence for cybersecurity questions at a point in time, but the relevance of those answers will be of unknown length.

While these challenges are real, none seem fundamentally at odds with the notion of empiricism or the scientific method. Indeed, a range of other disciplines faces one or more of these issues and still fruitfully make use of empiricism in practice. The common thread among all three of the problems is dynamism, which occurs in a variety of other disciplines as well. For example, while non-adversarial, insurers' assumptions about various kinds of property damage risk have been repeatedly updated and revised in response to observations that prior measured likelihood distributions were no longer predictive (e.g., concerning hurricanes or wildfires). Finally, closest to cybersecurity, economists routinely drive policy decisions using empirical tools, despite addressing a system that is, at its core, both adaptive and quasi-adversarial (few would seriously argue

---

[32] C. Herley and P.C. van Oorschot, 2017, "SoK: Science, Security and the Elusive Goal of Security as a Scientific Pursuit," *2017 IEEE Symposium on Security and Privacy (SP)* 99–120, https://oaklandsok.github.io/papers/herley2017.pdf.

[33] D. Evans and S. Stolfo, 2011, "Guest Editors' Introduction: The Science of Security," *IEEE Security & Privacy* 2011(9):16–17, https://www.computer.org/csdl/magazine/sp/2011/03/msp2011030016/13rRUwh80sR.

that the Federal Reserve would make better decisions if only it ignored empirical data). While each of these settings differs from cybersecurity in key ways, all are also removed from the idealized notion in which measurements can be used to derive static generalizable laws that then offer perfect predictive power. That such empirical analyses may be imperfect or have limited lifetime does not eliminate their value—for the alternative is to act without the benefit of concrete evidence at all.

In the late 20th century, portions of the medical community popularized evidence-based medicine (EBM) to incorporate a range of empirical evidence to guide research and ultimately update practice guidelines with the singular goal of improving clinical outcomes. Firmly embedded in the scientific method, EBM fostered hypothesis generation from laboratory and qualitative studies, building into both prospective and retrospective case studies, then driving repeated randomized controlled trials of prospective treatments and filtered based on comprehensive meta-analysis. There is little debate that this effort has been transformative for the practice of medicine. While this precise formulation is unlikely to translate directly to the cybersecurity realm, a similar kind of focus and investment to pursue outcome-focused results is needed.

However, there is a range of obstacles that will need to be addressed to make progress, including the following:

- *Data collection.* Evidence-based methods live or die based on the availability of evidence. Some kinds of analyses can be done in synthetic controlled settings, but much work requires in situ longitudinal data collection. Unfortunately, few organizations are organized around security-relevant data collection, and incentives—both due to privacy and liability—support holding such data closely. While it is straightforward to estimate how many Americans died of heart disease last year and their associated demographics, it is much harder to measure how many servers were compromised in each market sector, how many passwords were stolen from accounts in a position of authority, or the kind and number of new software vulnerabilities that were found and fixed and whether they were implicated in critical infrastructure (viz. Shostack's Cyber Public Health[34]). There are legal, operational, ethical, and economic aspects to this problem. How should such data collection be incentivized? What protections should exist for collectors, and how should the privacy interests of individuals implicated in the data be ensured? How might the use of such data be limited to improving cybersecurity? These are not simple challenges, and they implicate roles for the government, the private sector, and

---

[34] CyberGreen, "Indirect Cost Policy," https://cybergreen.net/technical-report-22-01/in, accessed February 6, 2025.

civil society. Improved cybersecurity data collection is not a problem that will be fixed in any single place, but it is "table stakes" for any significant progress.

- *Broadening analysis.* There is a tendency of cybersecurity discussions to focus on concrete artifacts—products and services. These are clearly important in any assessment, but it cannot be done at the expense of understanding what happens after such artifacts are fielded, and the processes and behaviors around how these artifacts are deployed and used. It is well understood, for example, that system misconfiguration—how a system is specialized for the environment in which it is operating—is as much of a security problem in practice as vulnerabilities in the underlying software systems. Similarly, most data breaches implicate some amount of social engineering (e.g., via phishing).[35] Systems are routinely developed to enforce a range of potential security policies, but there is only the most cursory understanding of what policies end up defined in practice, how and why they fail to capture the true security risks in their systems, and what solutions might better bridge gaps in IT staff knowledge and organizational behavior. As security is a holistic property, these inconvenient and hard to analyze realities cannot be ignored—from system administration to user training—and these are particularly places where empirical assessment will be critical.

- *Cultural change.* The computing community has been slow in embracing empirical and experimental methods as an approach for reasoning about cybersecurity. The core of computer science has traditionally been mathematical and mechanistic, and there can be a cultural tendency to strongly prefer solutions of that flavor. Indeed, when appropriate, such formal approaches are highly desirable as they can offer qualitatively comprehensive answers to scoped problems. Writing code in a language with a memory-safe type system (particularly one whose implementation and underlying runtime interface have been subject to adversarial empirical scrutiny) will generally be preferable to empirical studies about the past probability of memory errors in an unsafe language being detectable. However, outside of crisp properties such as memory safety, complex systems swiftly enter realms that are very much human-driven and with large numbers of hidden variables. This will require both the training of technologists in such methods and a cultural shift to recognize that such statistical deconstructive approaches (describing reality) are valuable and necessary complements to traditional strengths in more formal constructive ways of reasoning (defining reality).

---

[35] Verizon, 2024, "2024 Data Breach Investigations Report," https://www.verizon.com/business/resources/reports/2024-dbir-data-breach-investigations-report.pdf.

Despite these roadblocks, there has been considerable innovation in the "evidence-based security" space since the last version of this report. Among the approaches that have borne fruit are the following:

- *Opportunistic use of available data.* A community of researchers has been slowly building empirical analysis based on what data can be publicly gleaned. For example, open-source software revision control data, combined with bug databases and public repositories of vulnerabilities and exploits, have been used to identify a number of correlates of software vulnerability introduction, the efficacy of vulnerability-finding techniques such as fuzz testing, and the varied impact of "bug bounty" programs. Similarly, widespread active network scanning—enabled by tools such as zmap—and public breach reporting requirements have allowed correlating publicly visible security features with data breach outcomes.[36] The combination of "hard" (i.e., measurements) and "soft" (i.e., surveys) data is starting to emerge helping unravel thorny questions about the role of human behavior in security outcomes.
- *Outcome proxies.* Absent the ability to measure the impact of a security intervention directly, some research has explored the use of outcome proxies that capture some notion of adversarial hardness. For example, researchers have used the "underground price" of various criminal commodities and services to gauge the value of security interventions,[37] bulk shutdown of compromised accounts,[38] phone verification for account creation,[39] and account compromise.[40] The theory is that defenses that increase costs for attackers who offer their services at retail will naturally translate to an increase in their asking prices. In the same vein, others have offered bug bounty prices (both first party and from third parties such as Zerodium) as a metric to infer how hard it is to exploit particular popular platforms.
- *Second-order measurements.* Absent the ability to concretely measure security outcomes, a parallel line of work has focused on empirical assessments

---

[36] USENIX, 2015, "24th USENIX Security Symposium," August 12–14, https://www.usenix.org/conference/usenixsecurity15/technical-sessions/presentation/liu.

[37] A. Searles, Y. Nakatsuka, E. Ozturk, A. Paverd, G. Tsudik, and A. Enkoji, 2023, "An Empirical Study & Evaluation of Modern CAPTCHAs," pp. 3081–3097 in *32nd USENIX Security Symposium*, https://www.usenix.org/conference/usenixsecurity23/presentation/searles.

[38] K. Thomas, D. McCoy, C. Grier, A. Kolcz, and V. Paxson, 2013, "Trafficking Fraudulent Accounts: The Role of the Underground Market in Twitter Spam and Abuse," *22nd USENIX Security Symposium*, August 14–16, https://www.usenix.org/conference/usenixsecurity13/technical-sessions/paper/thomas.

[39] Computer and Communications Security, 2014, CCS '14: *Proceedings of the 2014 ACM SIGSAC Conference on Computer and Communications Security*, Association for Computing Machinery.

[40] A. Mirian, J. DeBlasio, S. Savage, G.M. Voelker, and K. Thomas, 2019, "Hack for Hire: Exploring the Emerging Market for Account Hijacking," pp. 1279–1289 in *WWW '19: The World Wide Web Conference*, May 13, arianamirian.com/docs/www2019_hfh.pdf.

of qualities that hopefully correlate with actual security outcomes. These include approaches such as process measurement,[41] relative measures of attack surface reduction,[42] or comparative analysis of bug-finding tools (e.g., fuzzers) on software corpora, as well as exploring the efficacy of transparency approaches such as a software bill of materials (SBOM).
- *Attackers and victims.* Multiple past and ongoing efforts have empirically explored concrete threat actors in great depth—for example, identifying attackers driven by social and financial reasons, or sponsored by nation states. Similarly, studies of victims and how the effectiveness of training, guidance, and nudges to reduce the likelihood of negative outcomes have led to significant improvements in major services.

Finally, one highly-desired manifestation of evidence-based security is the establishment of security metrics—parsimonious measures allowing the evaluation and/or comparison of the security offered by a particular system. The kinds of data that will be needed for an empirically based security research agenda are clearly amenable to being shaped into metrics. However, what makes metrics attractive is that they abstract and simplify. One can easily use metrics as a decision criterion for whether an organization has improved or not, whether vendor A or vendor B is more secure, or as defense against liability. However, this same attractiveness creates strong incentives for standardization and can create institutional inertia that makes it difficult to change or react when the context in which the metrics were measured has changed. Even worse, failures in analysis or validation might elevate "bad" metrics, which ultimately incentivize less secure decisions. The same result can come from the incentive to "game" good metrics. Thus, it is critical to also consider how to deliver evidence-based security research, without enabling the most negative aspects of institutional desires for cheap decision making.

Many security metrics, from "speed to patch" to the collection of end user agent parameters, have been developed. However, there is no basis for believing any set of existing metrics provides an accurate prediction of safety or a root-cause analysis of previous losses.

While there is some research on software developers' ability to produce secure software, this research seems to have had little impact on actual software development tools and processes. Secure development life-cycle practices and the tools that support them have largely come from industry and often have their own usability challenges.

---

[41] Cybersecurity and Infrastructure Security Agency (CISA), 2016, "Alert: OpenSSL 'Heartbleed' Vulnerability (CVE-2014-0160)," October 5, https://www.cisa.gov/news-events/alerts/2014/04/08/openssl-heartbleed-vulnerability-cve-2014-0160.

[42] Github, "AttackSufaceAnalzyer," Microsoft, https://github.com/microsoft/AttackSurfaceAnalyzer/pulls, accessed February 6, 2025.

Some training to help software developers produce more security software can produce recommendations that seem on the surface impractical.

The inability to develop compact, predictive, measurable security metrics informs and affects essentially all of the consumer cyber hard problems.

Secure Resilient Engineering and Formal Methods for High Assurance

Many of the traditional problems of "secure design and composition," described in detail above, include careful specification, isolation, and partitioning of functionality—following the principle of "least privilege" by authenticating the principal on whose behalf actions are taking and verifying their "right" to take such an action (the basis for "zero trust"). There are also emerging technologies, such as "confidential computing," which provides a strong, principled basis for authentication of programs (an important security principle) to establish a principled, distributed basis for partitioning, isolation, and trust management. Several "producer technologies" for achieving secure resilient design and composition are discussed below.

There is a wide diversity of engineering interventions, ranging from selecting safer programming languages such as Rust and TypeScript to making architectural choices that enhance the potential for resilient response to compromise. As noted above in this chapter, one of the most vexing cyber hard problems in cybersecurity is measurement. Measurement difficulties thwart progress, not just in implementing secure engineering practices but also in understanding trade-offs when there are choices to be made regarding which practices to adopt.

For example, how much benefit is to be obtained from using a safer programming language, and how does this compare, say, with using improved analysis tools? There may be trade-offs that involve traditional software engineering criteria—for example, how does architecting to reduce interdependency among components of a large system interact with choices to implement a design that limits trust assumptions among system components when they interact (in the same sense as zero trust, but at an internal implementation level)?

Further exacerbating this challenge is a set of enduring perceptions that engineering efforts that are directed at enhancing security and resilience have an uncertain return on investment.

When successful, however, secure engineering practice can have significant—and otherwise unattainable—benefits in reducing many aspects of cyber risk. Indeed, while it may be challenging to characterize the role of specific development or design practices, there is a range of proxies suggesting that "zero-day" exploits for commodity smartphone and server platforms have become harder to procure over time, as evidenced by their increasing market price and owing to the increasing length and complexity of exploit chains needed. Moreover, there are cases where secure engineering practice leads not only to improved risk posture but also improved productivity and, in some instances, enhanced system performance. It is reported, for example, that many users of memory-safe programming languages (e.g., Rust and TypeScript) make the adoption choice in the interests of productivity, with secondary consideration for the harder-to-measure security benefits. This means that, even if the security benefits cannot readily be measured, the adoption of improved practices based on concomitant benefits to productivity and performance can nonetheless be promoted.

There are many examples of guidance regarding secure engineering from firms, laboratories, and government agencies, including the National Security Agency, the Cybersecurity and Infrastructure Security Agency, and ONCD. Most recently, for example, ONCD issued advice regarding secure practices with a focus on memory safety and formal methods[43] in the face of the paucity of cybersecurity metrics. Historically, the Microsoft Security Development Lifecycle (SDLC), which includes a mix of interventions and practices focused on process and product, has had broad adoption and, as perceived by engineering managers, meaningful benefits.[44] Much of the guidance focuses on practices and processes that are associated with improved security outcomes. An example is "secure coding practice," which involves making coding choices that reduce vulnerabilities, as demonstrated through techniques such as fuzz testing.[45]

Logic Flaws and the Need for Mathematical Techniques

Many security-related weaknesses and vulnerabilities go beyond simple coding practices and derive from the logical structure of software and firmware. These flaws can range from protocol and API misuse to erroneous business rules. These are logic flaws. The focus of logic flaw problems is not on the full scope of secure engineering practices, but rather on means to achieve verifiable assurances regarding the absence of certain categories of vulnerabilities that go beyond type safety and memory safety. This report focuses on logic flaws for two reasons. First, there is increasingly broad adoption of "traditional" secure engineering practices such as SDLC (and as assessed through instruments such as BSIMM).[46] Second, logic flaws account for an increasing percentage of exploits,[47,48] and these flaws are not readily detected using current techniques. The usual means to detect

---

[43] ONCD, 2024, *Back to the Building Blocks*.
[44] Microsoft, "Microsoft Security Development Lifecycle (SDL)," https://www.microsoft.com/en-us/securityengineering/sdl, accessed February 6, 2025.
[45] OWASP Foundation, "Secure Coding Practices," https://owasp.org/www-project-secure-coding-practices-quick-reference-guide/stable-en/02-checklist/05-checklist, accessed February 6, 2025.
[46] BlackDuck, "What Is BSIMM?" https://www.blackduck.com/glossary/what-is-bsimm.html, accessed February 6, 2025.
[47] OWASP, "Top Ten," https://owasp.org/www-project-top-ten, accessed March 25, 2025.
[48] S. McClure, 2024, "Safeguarding from Lurking Threats in Business Logic Flaws," *Fast Company*, January, https://www.fastcompany.com/91013667/safeguarding-from-lurking-threats-in-business-logic-flaws.

and mitigate logic flaws is through inspection and testing. But it is well understood that these methods are imperfect.

There is an unhappily rich collection of security attributes, as revealed in the several taxonomies that are widely referenced (e.g., CIA,[49] STRIDE,[50] MITRE ATT&CK). Mathematical techniques can be used to address some of these, at different levels of complexity and scale. However, some attributes cannot currently be readily modeled mathematically. These include side-channel attacks. An example familiar to security researchers is cryptographic algorithms, which are mathematically correct and whose implementation is proved consistent with the algorithms, but whose implementations on actual physical processors creates vulnerabilities based on the physics of the operation of the processors, such as power fluctuations, RF emissions, and timing of executions.

Process compliance, in many cases, is seen as more affordable to achieve than actual measurable security. Although sometimes useful,[51] compliance is expensive, subject to manipulation by "well-resourced" organizations, and it delays innovation and is not really very effective, generally, in "guaranteeing" security.

Formal methods (FM), in contrast, are direct techniques, focused on the operation rather than adherence to processes in the creation of that product. These techniques, including verification and program analysis for various functional and quality attributes, have a long history, going back at least to the 1960s with work by Robert Floyd and later Tony Hoare. For many years, the principal uses were in critical applications such as commercial flight controls, embedded medical devices, and national security applications. In the past 5 years, however, the scope of application has broadened significantly to include many commercial uses. Some evidence of this is cited in the Networking and Information Technology Research and Development publication regarding the FM@Scale workshops,[52] where several at-scale commercial uses are highlighted. These use cases suggest the possibility that barriers of affordability, scale, usability, and integration can be overcome for a broader range of applications, with significant benefits not just to reducing important categories of vulnerabilities but also in providing evidence in support of cybersecurity risk assessment and certification. These successes depend on our ability to express models, including specifications, for quality attributes relevant to security. Improving the scope, expressiveness, and ease of use of modeling can significantly

---

[49] CIA refers to confidentiality, integrity, availability.

[50] STRIDE refers to spoofing, tampering, repudiation, information disclosure, denial of service, elevation of privilege.

[51] The National Information Assurance Partnership (NIAP) Common Criteria, for example, can involve deep analysis of design artifacts and some sampling of code in an evaluated system. See NIAP, "Common Criteria: IT Security Evaluation," https://www.nsa.gov/Portals/75/documents/resources/everyone/2023-02-NIAP_brochure_trifold_1.pdf, accessed February 6, 2025.

[52] R.W. Floyd, 1967, "Assigning Meanings to Programs," pp. 19–32 in *Proceedings of Symposium on Applied Mathematics (19): Mathematical Aspects of Computer Science*, J.T. Schwartz, ed., American Mathematical Society.

reduce the cycle of information loss and recovery that typically occurs, with great cost, as systems evolve over time. Information loss of this kind creates challenges for test and evaluation as well as for sustainment and evolution. It is one of the chief contributors to technical debt.[53]

An additional enabler, already in place, for maintaining continuity, enabling agility, and minimizing information loss across the full life cycle of software-based systems is the Software Acquisition Pathway. The benefits were reinforced in a recent memo from the Secretary of Defense.[54,55]

The following three examples illustrate the various ways that these barriers are now being overcome:

- Defense Advanced Research Projects Agency's (DARPA's) High Assurance Cyber Military Systems program, circa 2012–2017 (and recently awarded a DARPA Game Changer Award), undertook case study exercises to assess and demonstrate the potential applications of FM techniques to mid-sized complex systems, including unmanned aerial vehicles. The approach to scaling is significant: Judicious choices regarding systems architecture and software infrastructure enabled a narrowing of focus in use of FM to a relatively small portion of the system, including the underlying operating system, seL4, and adherence to the key coupling constraints imposed by the systems architecture.[56,57]
- One of the most significant FM interventions is the use of types—and more recently memory-safety features—in programming languages. Type declarations form a specification, and type checking verifies that the software is consistent with the declarations. The declarations are just another element of the language specification, and the type-checking algorithms, which are mathematically complex, are integrated into the compiler. Adding to the ease of use is type inference, where types can be determined algorithmically, enabling specifications to be largely omitted. The mathematical analysis, nearly fully invisible to users, can eliminate entire classes of vulnerabilities related to

---

[53] CISA, 2025, "Closing the Software Understanding Gap," January 16, https://www.cisa.gov/resources-tools/resources/closing-software-understanding-gap.

[54] DoD, 2025, "Directing Modern Software Acquisition to Maximize Lethality," Memorandum for Senior Pentagon Leadership Commanders of Combatant Commands Defense Agency and DoD Field Directors, from the Secretary of Defense, March 6, https://media.defense.gov/2025/Mar/07/2003662943/-1/-1/1/DIRECTING-MODERN-SOFTWARE-ACQUISITION-TO-MAXIMIZE-LETHALITY.pdf.

[55] DoD, 2020, "Operation of the Software Acquisition Pathway," DoD Instruction 5000.87, October 2, https://www.esd.whs.mil/Portals/54/Documents/DD/issuances/dodi/500087p.pdf.

[56] Defense Advanced Research Projects Agency (DARPA), "HACMS: High-Assurance Cyber Military Systems," https://www.darpa.mil/program/high-assurance-cyber-military-systems, accessed February 6, 2025.

[57] A. Miller, "HACMS," GALOIS, https://galois.com/project/hacms-high-assurance-cyber-military-systems, accessed February 6, 2025.

consistency of interpretation of low-level representations and, with languages such as Rust, related to the safe mutation of shared data.[58]

- The Amazon Web Services (AWS) cloud service is a highly complex distributed system that could manifest diverse errors related to concurrency. One of the challenges of concurrency is that errors are intermittent and at a potentially low rate. This means that testing is unlikely to find the contributing faults in the code, but that at-scale execution in a distributed cloud could nonetheless manifest errors at high frequency. Using TLA+, AWS engineers have been able to find subtle design bugs and then, on repair, prove absence of similar design bugs.[59] There are many other applications. One example is the use of FM to support verification and querying of AWS security access policy implementations using satisfiability modulo theories (SMT) provers; the provers run billions of times each day to support this verification.[60,61]

It is evident that there is progress in FM and that barriers to its use can be overcome in specific cases, including usability, scale and composition, and integration into engineering workflows. These examples are illustrative, but there are significant barriers.[62] The recent ONCD report highlights two areas—software and hardware memory safety, such as provided through Rust and similar languages, and FM to support affirmative claims—backed by evidence—regarding particular security properties.[63]

The solution of these problems and application and further development of these techniques acutely influences hard problems 1, 2, and 3 but also 8, 9, and 10.

Cyber-Physical Systems

A recent President's Council of Advisors on Science and Technology (PCAST) report[64] drew attention to critical CPS (which they call "operational technology systems"). These systems include the "integrated digital and physical resources that are crucial to

---

[58] C. Thompson, 2023, "How Rust Went from a Side Project to the World's Most-Loved Programming Language," *MIT Technology Review*, February 14, https://www.technologyreview.com/2023/02/14/1067869/rust-worlds-fastest-growing-programming-language.

[59] L. Lamport, 2024, "The TLA+ Home Page," August 13, https://lamport.azurewebsites.net/tla/tla.html.

[60] Amazon Web Services, "Provable Security Resources," Cloud Security, https://aws.amazon.com/security/provable-security/resources, accessed February 6, 2025.

[61] N. Rungta, 2022, "A Billion SMT Queries a Day," *Amazon Science*, https://www.amazon.science/blog/a-billion-smt-queries-a-day.

[62] P. Lincoln, W. Scherlis, and W. Martin, 2022, *Formal Methods at Scale: 2019 Workshops Report*, Computing-Enabled Networked Physical Systems Interagency Working Group, May, https://www.nitrd.gov/pubs/Formal-Methods-at-Scale-Workshops-Report.pdf.

[63] ONCD, 2024, *Back to the Building Blocks*.

[64] President's Council of Advisors on Science and Technology (PCAST), 2024, *Strategy for Cyber-Physical Resilience: Fortifying Our Critical Infrastructure for a Digital World*, Executive Office of the President, February, https://bidenwhitehouse.archives.gov/wp-content/uploads/2024/02/PCAST_Cyber-Physical-Resilience-Report_Feb2024.pdf.

Americans' daily lives, including the electrical grid, public water systems, internet and telecommunications, air traffic control and much more." IoT and automated manufacturing systems that incorporate sensors or actuators, and indeed any CPS, can be included in this category as well. The PCAST report goes on to note:

> Cyber-physical risk is high, while protections are disproportionately low. America's infrastructure systems were created and operated long before they acquired cyber dependencies, with sensing, computing, and networking dependencies developing in different ways over time. There is no systemic, pervasive protection against cyber risk since our protections evolved over time.[65]

CPS (e.g., a car, a laboratory instrument, or a medical device in a doctor's office) may have very long lifetimes. Future systems have to be shaped by cyber-informed engineering. Much of the technology that underpins cyber systems and CPS was engineered without appropriate consideration of security needs. Security and resilience elements are tacked on after systems are deployed, often imperfectly and at considerable expense.

As observed above, many CPS security practices lag far behind those of IT systems. These systems use old software versions and are often not upgraded, they are not managed under user policy, and there is seldom support for critical hardware features like "root of trust" nodes. Finally, they are especially vulnerable since they are exposed to physical attacks in addition to network-based attacks.

This is cyber hard problem 8, but it is affected by almost all the other "producer" cyber hard problems, including problem 10 (operational security).

OPERATIONAL CYBER HARD PROBLEMS

Operational cyber hard problems address securely operating a scale infrastructure, including responding to attacks. This is cyber hard problem 10. Earlier discussion has already described the importance of continuous updating, resilient deployment and operations, monitoring, and situational awareness. These are the core elements of resilient secure operation, and large cloud providers have made great progress in this area. However, customer insight into the effectiveness of already introduced measures is modest, and small providers as well as "in house operations" often suffer in comparison to the effectiveness of the operational security of a large cloud provider. This includes the operational infrastructure of CPS.

---

[65] PCAST, 2024, *Strategy for Cyber-Physical Resilience*, p. 12.

Understanding Human Attributes of Attackers

It is important to acknowledge that there are human attackers who make decisions and can be influenced to defenders' advantage. These attackers have a profound influence on the secure operation of a cyber system.

For much of the history of cybersecurity, defenders have thought about attackers as an abstract and homogenous group. This meant that defenses were applied to technical attributes of attacks, such as techniques and tools, rather than through understanding the humans behind the attacks. Maturing threat models identifying threat groups can differentiate capabilities, objectives, and likely victims. A cyber hard problem, however, remains incorporating insights about the human attributes of attackers to create tailored defenses.

Attackers routinely exploit the human weaknesses of their victims; however, defenders lack sufficient insights to effectively incorporate human factors into defense. Historically, deception, such as honeypots, has been used to manipulate attackers to gain intelligence, but these techniques are neither widespread nor generally evaluated for effectiveness. In addition, it is difficult to persuade engineering organizations, which are mainly rewarded for adding new features to work on adversarial engineering projects. Recent research on adversarial human factors is starting to identify and analyze the human attributes of attackers. More research funding is needed. Industry partnership will likely be critical for developing commercial defensive capabilities that apply adversary human factors. Together, industry and the research community have to develop metrics for evaluating the effectiveness of the approach.

Understanding the human attributes of attackers is crucial for several reasons. Primarily, it allows for more sophisticated and targeted defense mechanisms that go beyond merely blocking attacks or deactivating users to anticipating and mitigating them. This knowledge can significantly improve the ability to prevent breaches and reduce the impact of successful attacks. Governments, businesses, and individuals all have a stake in this issue because the consequences of cyberattacks can range from financial loss and reputational damage to national security threats. By understanding the motivations, psychological traits, economic incentives, and behavioral patterns of attackers, defenders can craft strategies that are more likely to deter or disrupt malicious activities.

The complexity of this problem lies in the inherent variability and adaptability of human behavior. Attackers come from diverse backgrounds, possess different skill levels, and are driven by a range of motives, including financial gain, political activism, espionage, or personal vendettas. The dynamic nature of human attributes makes it challenging to create a one-size-fits-all defense. Moreover, ethical and privacy considerations must be balanced when researching and using human factors in cybersecurity.

Lawyers taking a position that even attackers have privacy rights limits defender access to their internal communications and communication with fraud victims. The incentives for attackers are high and constantly evolving, while defenders often face significant resource constraints and organizational inertia, making it difficult to implement and adapt new strategies to increase the cost seen by attackers.

Potential approaches to this problem include interdisciplinary research that combines insights from psychology, sociology, and behavioral economics with cybersecurity. Developing comprehensive attacker profiles and predictive models can help in understanding likely behaviors and vulnerabilities. Simulation and gaming techniques can be employed to study attacker behavior in controlled environments. Additionally, leveraging machine learning (ML) and AI to analyze patterns in large data sets can provide deeper insights into attacker traits and tactics. Collaboration between academia, industry, and government will be essential to foster innovation and translate research findings into practical applications. Note, however, that attackers will be increasingly autonomous (bots) and will also employ AI to probe systems and people.

Key players who can act include cybersecurity researchers, defense agencies, technology companies, and policy makers. Cybersecurity firms and tech companies can incorporate human factors insights into their security products and services. Governments can fund research initiatives and create frameworks that encourage information sharing and collaboration. Policy makers can help by enacting regulations that support ethical research, while researchers can make progress in blinding and masking personal information to make it less easy to reidentify, to protect individual privacy. Education and training programs for cybersecurity professionals need to also emphasize the importance of understanding attacker psychology and behavior.

Success in this endeavor can be measured through several indicators. A reduction in the frequency and severity of successful cyberattacks would be a primary indicator. Improved response times and more effective mitigation strategies in the face of attacks would also signify progress. An increase in the cost of credentials or breach data offered on dark markets would be a direct observable measure of success. Additionally, the development and widespread adoption of new defense technologies and methodologies that incorporate human attributes of attackers would demonstrate advancement in this field. Regular assessments and refinements based on real-world data and feedback will be necessary to ensure the ongoing effectiveness of these approaches.

This informs cyber hard problem 10.

Personal and Societal Impacts in Design and Operation

There are few tools or techniques to predict how cyber systems may be used in the future, especially as relates to unforeseen use that can threaten an individual's security, the security of an organization, or the nation. Attacks that seem to have been largely unforeseen in the past include doxing, stalkers transitioning to the digital domain, false reporting to cause account deactivation, and disinformation and influence operations that aim to attack democratic processes.

The legacy of online anonymity, including the ease of creation of opaque personas in different forums, facilitates not just free speech but also an important category of adversarial information operations, the use of bots, artificial amplification of adverse memes, and impersonation of individuals and organizations. In certain online contexts, by contrast, identity has to be firmly established, such as for banking, health records access, and business process execution.

The challenge is to identify multifaceted approaches to identity that could thwart the adversarial behaviors that depend on lack of provenance while nonetheless supporting the many contexts where it is needed, including free speech and augmenting protections of the identity of victims and potential targets of abuse and attack (e.g., human rights workers).

Anonymity also facilitates socially destructive behaviors such as stalking, bullying, sexual harassment, and doxing (piercing the veil of the privacy of other people). What are technical approaches or potential remedies that do not require abandoning the free-speech benefits of anonymity?

Human-Centered Design of all Human-in-the-Loop Security-Related Interactions

Computer users are called on to make decisions and choices with cybersecurity consequences that they do not have sufficient knowledge and resources to make accurately or securely.

Decisions and choices may be explicit or implicit. Users may be using their own computer or a computer owned by their organization. They may be acting as an individual or as a member of an organization, such as an employee or even an administrator. They may be using special purpose tools as experts in a particular discipline (i.e., code development or threat hunting). The resources they lack to address the decisions thrust upon them, some of which may be irrevocable, are legion—information, knowledge, understanding, context, time, memory, attention, desire, and incentive.

The Verizon *2024 Data Breach Investigations Report*[66] calls out that approximately two-thirds of breaches involve a (non-malicious) human element. This reflects the tendency, as noted in the IAM discussion above, for all sorts of cyber products to "kick the can down the road" to required human decision making, when the humans involved do not have sufficient resources to do so. In the IAM case above, what is lacking is

---

[66] Verizon, 2024, *2024 Data Breach Investigations Report*.

transparency into the system and an understanding of the system's complexity, which seems obviously like a too-heavy lift to require of a user.

Even in the most straightforward context, research on expert advice about what security practices users, as consumers, need to follow shows that experts do not agree on the most important.[67] The amount of expert advice is overwhelming and not necessarily coherent.

The responsibility for users to make impossible security decisions creates vulnerabilities that attackers discover and exploit using social engineering. The security impact from such attacks may be to the users themselves, their computers, their organization, or to the Internet at large. The unmet expectations placed on users deteriorate presumed protection levels (which may go undiscovered until it is too late). This can contribute to a state of habituation, or worse, learned-helplessness, for users—Why follow the security advice if I am not able to do it all, or even know which matters?

Testing with appropriately representative users is both expensive and difficult. There are no tools, frameworks, principles, or automation to do the required testing without humans in an adequate and rigorous fashion. There are still gaps between the best-of-breed testing methodologies in research and what's available to testers, who are often entry-level employees.

While the techniques for testing are largely known, choice of testing subject needs to consider the user's skills, knowledge, and context. Since those vary, testing needs to include a range of such subjects. In addition, because some of the cybersecurity-related decisions may come about from an error in the system, or an attack, the suite of interactions needing testing can be difficult to identify, and difficult to replicate for testing.

When users interact with devices and systems in unmanaged environments (personal use), they acquire ad hoc habits that are ineffective or insecure, and these are difficult to unlearn. Because there is often no discernable difference between effective security advice and ineffective security habits, users become accustomed to ignoring security advice and experiencing no known harms.

Builders may "punt to the user" to make decisions in uncertain design situations, which compounds the issue at hand. This shows up as either warnings presenting choices or new configuration settings, which are often hidden. If used correctly (with omniscience), perhaps the security would be improved, however usability of said features goes untested.

Security and privacy mechanisms today reflect current knowledge on how to expose and control, which often differs from the user's cognitive model of the system.

---

[67] E.M. Redmiles, N. Warford, A. Jayanti, A. Koneru, S. Kross, M. Morales, R. Stevens, and M.L. Mazurek, 2020, "A Comprehensive Quality Evaluation of Security and Privacy Advice on the Web," *Proceedings of the 29th USENIX Security Symposium*, August 12–14, https://www.usenix.org/system/files/sec20-redmiles.pdf.

Builders build what they know how to build. Users use what they are given. The resulting gap present a symmetry of ignorance that must be overcome.[68]

Helping Security Professionals Help Security

This is part of cyber hard problem 6, "human–system interactions," and cyber hard problem 2, "secure development."

Even if human-centered design were addressed, support for human security workers is still insufficient to ensure optimal security outcomes. The workforce is core to cybersecurity, in practice, but there is not necessary support in place to help security workers thrive and perform, with education, tools, and processes. This includes workers of all sorts, including developers, designers, architects, IT, and specialists such as security architects, chief security officers, blue teams, content moderators, and fact checkers.

There are two interconnected facets to this problem—inadequate training and the inability to provide cybersecurity without that support. For generalists who are also security workers, the support needed includes training, resources, and incentives baked into the job, not exogenous. Education available to learn to code would teach best practices for coding securely. Specialized security worker education would cover the myriad of topics needed to be a security subject-matter expert—from architecture, to design, to coding security functions, to specialized security testing, to security in deployment and use.[69]

The issue of inadequate tools and techniques for security workers to develop secure code is addressed in the cyber hard problems above.

The pool of trained security workers is limited, and training in one job can lead to the ability to move to a better paying job, creating an ongoing training need. Jobs that involve operational security vigilance are high pressure and can lead to burnout. Lack of certification means it is impossible to enforce education, training, and standards of professional conduct.

Since the 1996 New Security Paradigms Workshop,[70] there has been substantial work, both in research and practice, on usable security and privacy, particularly for individuals and consumers. However, there is less work on the many other humans involved in creating, maintaining, operating, receiving, and even attacking, security and privacy, and how they can be supported, or repelled.

---

[68] O. Pieczul, S. Foley, and M.E. Zurko, 2017, "Developer-Centered Security and the Symmetry of Ignorance," *Proceedings of the 2017 New Security Paradigms Workshop*.

[69] As examples, NIST and the SANS Institute have training materials available for all levels from end users to cybersecurity professionals. See NIST, "Free and Low Cost Online Cybersecurity Learning Content," Applied Cybersecurity Division, https://www.nist.gov/itl/applied-cybersecurity/nice/resources/online-learning-content, accessed February 6, 2025.

[70] M.E. Zurko and R.T. Simon, 1996, "User-Centered Security," New Security Paradigms Workshop, https://www.nspw.org/papers/1996/nspw1996-zurko.pdf.

Research in practical secure deployment is even sparser and tends to focus on individuals, largely overlooking organizational issues. Some research exists on why, how, and when individuals will accept security-related updates. Studies and measurements exist showing the rate of patch deployment. Mistakes are made, and attacks take advantage of those gaps. Research on expert advice about what security practices to follow shows that experts do not agree on the most important.[71]

Designing for Resilience

Resilience is more difficult to retrofit into existing systems as an afterthought. It requires thoughtful architectural design from the outset, considering factors such as graceful degradation and partitioning of mission-critical functions to minimize the impact of breaches. The goal is for systems to operate securely, albeit in degraded fashion, even when some components are compromised. This attribute is increasingly important as large-scale systems are interconnected into even larger-scale systems, with the larger goal of organizational resilience and the ability to operate essential business functions even when systems are impaired.[72]

The inherent uncertainty of cyberattacks often leads companies to delay investments despite the broader societal benefits of resilient systems. Therefore, integrating resilience into the initial design phase is far more effective than attempting to retrofit it, highlighting the importance of preparedness and proactive planning. In addition to the benefit of training, resilient design is still a matter of active research, especially regarding measuring the resilience properties of a composed system.

As cyber threats evolve, the need for resilience in maintaining operational continuity becomes increasingly urgent. Resilience is the key to ensuring that vital services, such as health care, finance, and critical infrastructure, can withstand and recover from attacks. This issue is paramount to the broader community, including businesses, consumers, and governmental bodies, as it directly impacts economic stability, public safety, and national security.

Potential approaches to enhancing resilience include both technical and organizational strategies. Again, as mentioned earlier, adopting architectural choices that support resilience in designs, including for distributed systems, as well as architecting for minimal trusted computing bases, can be beneficial. Formal or semi-formal verification of design and implementation can ensure that critical systems meet high standards of resilience.

On an organizational level, fostering a culture of preparedness and operational excellence is essential. This includes thorough risk assessments, continuous improvement

---

[71] E.M. Redmiles, N. Warford, A. Jayanti, A. Koneru, S. Kross, M. Morales, R. Stevens, and M.L. Mazurek, 2020, "A Comprehensive Quality Evaluation of Security and Privacy Advice on the Web," *Proceedings of the 29th USENIX Security Symposium*, August 12–14, https://www.usenix.org/system/files/sec20-redmiles.pdf.

[72] CISA, "Secure by Design," https://www.cisa.gov/securebydesign, accessed February 6, 2025.

of security configurations, and minimizing the "blast radius" of potential attacks by effectively partitioning critical and non-critical functions. Information-sharing models, akin to those used by the Federal Aviation Administration or anti-spam initiatives, can enhance collective resilience by disseminating threat intelligence and best practices across organizations. Finally, preparation requires deliberate practice to ensure that people and plans are effective, agile, and ready to respond to real-world incidents.

Advancing resilience is a collective effort that involves multiple stakeholders. Government agencies play a crucial role by setting standards and incentivizing resilient design, while industry groups can contribute by developing and promoting best practices. Companies, especially those operating critical infrastructure, have to make resilience a priority in their design and operational processes. It is important that resilience strategies be practical, scalable, and widely adopted.[73]

Various indicators can be used to "measure" success in designing for resilience. For example, these include the system's ability to maintain functionality during attacks, local and network outages and natural disasters, the speed and effectiveness of recovery processes, and the overall reduction in the impact of cyber incidents. Regular testing, simulation of attack scenarios, and continuous improvement based on feedback and threat intelligence will help gauge progress. Ultimately, a resilient cybersecurity posture will not only mitigate the damage from attacks but also instill greater confidence in the security and reliability of all systems.

This is cyber hard problems 1, 2, and 3, as well as its effect on problem 10.

Situational Awareness of Defenders

Growing software, system, and network size, complexity, and usage offer attackers increasing opportunities for both successful penetration (i.e., larger attack surface) and the ability to remain undetected and operate within the compromised environment (i.e., larger persistence volume). The former concern is generally addressed through secure system design and implementation, while the latter is addressed by intrusion detection and digital forensics. Although some progress has been made in software and system hardening (at least against certain classes of easier-to-exploit vulnerabilities), it appears that the dwell time of non-ransomware-focused sophisticated attackers remains high,[74] despite significant investment in the collection, monitoring, and analysis of security-relevant events. Essentially, the duality of detection and evasion in cybersecurity

---

[73] The Global Resilience Federation, a nonprofit, offers framework concepts for operational resilience for business. See Global Resilience Federation, "The Operational Resilience Framework," https://www.grf.org/orf, accessed February 6, 2025.

[74] The committee excludes ransomware because it inherently exhibits a very obvious and "noisy" behavior relatively soon after infection. Examples of more recent, long-dwell threat actors include TRIANGULATION and Volt Typhoon.

continues to trend in ways favorable to the sufficiently motivated and well-resourced attacker.

As a result, successful defense and remediation requires timely identification and disruption of malicious activities, whether proactively or reactively. The current state of practice relegates defenders to playing "whack-a-mole" while peering through a keyhole, with in-band network management tools that are potentially influenced by the attacker. The current cybersecurity tools and practices are not sufficiently precise to reliably identify the activities of slow-and-stealthy attackers over extended periods of time, nor efficient or fast enough to identify and stop rapid-moving attacks. Furthermore, a lot of emphasis has been placed on exfiltration detection, with significantly less on detecting other types of cyberattacks such as scheduled system-level denial of service.

Several technical factors contribute to confounding the ability of defenders to achieve a sufficient level of situational awareness. These include, but are not limited to, (1) improved threat actor tactics (low "signal") that increasingly take advantage of native features and resources[75] of the targeted environment; (2) high volume of benign system events (high "noise") as a function of system size and complexity; (3) high volume of low-sophistication attack events (high "background radiation") that lead to alert fatigue and misprioritization of response resources; (4) more complex and diverse system capabilities that, at least from an observability perspective, partially overlap with attacker capabilities, objectives, and behaviors (e.g., built-in screen recording, system-wide document search); (5) continued reliance on human-driven analysis (threat hunting), with the corresponding limitations on volume and pace of analysis; (6) the inability of analytics to keep up with ever-growing, security-driven telemetry data volumes; and (7) diminishing returns (low "gain") but high, continuous, fixed cost for any additional type of telemetry collected and used, due to the rich set of pathways attackers can exploit to meet a given objective. Sociotechnical factors that also negatively contribute to the problem include the high cost for security-data storage and processing, the friction of information and data sharing and analysis across intra-organizational boundaries, lack of trained personnel, and (a sometimes real) conflict with other legal, privacy, or regulatory requirements.

No single solution appears sufficiently powerful to fully or substantively address the detection problem on its own. However, with appropriate investment for further scientific investigation, the following practices and technologies could play a positive role in addressing the problem:

- Software, system, and network partitioning[76] and tailored system monitoring can increase observability and improve attack detection, if actually analyzed

---

[75] This practice is often referred to as "living off the land."
[76] Such partitioning appears to offer several security benefits, at the potential cost of overall complexity.

(and so carefully engineered to avoid data overload). What is needed are solid principles and tools for enabling system architects, implementers, and (critically) administrators to design, build, and manage such deployments.

- Certain types of system structure are thought to make change detection and anomaly detection easier. For example, it is anecdotally believed that certain types of cloud computing offerings (in particular, software-as-a-service environments) are easier to defend due to the uniformity of the normal behavior of the computing and software infrastructure. Past government-sponsored efforts, such as the DARPA Transparent Computing program,[77] explored aspects of system and analytic co-design for observability and cybersecurity reasoning leveraging the concept of determinism. What is needed are further studies, tools, and methodologies supporting data- and metric-driven decisions with respect to designing and implementing complex systems for observability.

- The application of AI to the problem of information summarization, knowledge extraction, and presentation and interaction—tailored to the problem of situational awareness—offers the potential of reducing the increasing volume of telemetry to concise, contextualized, understandable, and actionable material. Existing work, such as MITRE ATT&CK,[78] provides an initial framework for organizing information and knowledge, but further work is needed for determining how best to use AI in this setting.

- Beyond direct enablement of human analysts, AI could address the problem of high cost and diminishing returns of security telemetry by continuously and dynamically orchestrating, evaluating, and acting on selective telemetry sources in response to updated models of hypothesized attacker activity based on prior observables.

- The use of AI-driven agents for conducting large-scale red teaming can help with human analyst training; better tuning of detection models; identification of blind spots; and, at a strategic level, going beyond anecdote-driven security practices toward using better global metrics on the effectiveness of the various technologies and approaches in the field.

Most of the above practices and technologies would be significantly aided by the availability of open, high-fidelity, experimental test beds. These would need to go beyond the typical goal of offering representative topologies, systems, and software to include realistic (ideally real) background data and activity sufficient to simulate actual environments and scenarios.

---

[77] DARPA, 2014, "Transparent Computing," https://www.darpa.mil/research/programs/transparent-computing.
[78] MITRE, "ATT&CK," https://attack.mitre.org, accessed February 6, 2025.

The fundamental and second-order metrics for success[79] remain relevant and appropriate for evaluating individual technologies and, in some cases, combinations thereof. The primary challenges that need to be addressed are (1) providing higher-quality and unbiased evaluation of these technologies[80] that goes beyond sparse, anecdotal empirical testing (e.g., human-driven red teaming) and (2) translating specific-technology in-lab effectiveness measurements to real-world impact. With respect to the latter, when better and consistent metrics of effectiveness become the norm, sharing of system and network security architecture patterns along with measurements would go a long way toward establishing a proper engineering discipline in this space.

Damage Assessment and System Reconstitution

Accurately determining the initial vector and subsequent impact of a cyberattack has always been a time-consuming and difficult task. A complete analysis would identify several actionable aspects of the attack, including the method of compromise; the software, systems, and users through which initial infection occurred; the full set of systems and data accessed, exfiltrated, or modified by the attacker; and any new software introduced, existing software modified, configuration changes made, or upstream and downstream services accessed during the attack—while at the same time filtering out the typically much larger volume of benign, legitimate activities that may be overlapping and interleaved with attacker activities in both time and space (i.e., in the same systems during the same time period). These are necessary for determining how an attacker was able to gain initial access (to prevent reinfection), what data were lost (e.g., to determine what intellectual property or customer personally identifiable information [PII] was stolen), what assets (e.g., critical infrastructure components) were tampered with, whether the attacker has been completely evicted, what latent access vectors an attacker may have introduced (again, to prevent reinfection), and what residual risk must be dealt with (and potentially through what methods). In many cases, the full extent of the damage incurred is often revealed only after significant time has elapsed since the initiation or even the discovery of the attack. Relying on reported extrinsic observables (e.g., reported financial fraud, identified damage on devices, or cyber-physical processes) negates much of the potential for timely intervention and prevention (or at least minimization) of said damage. For high-stakes events, teams of specialist forensic analysts must work manually over several weeks or months to produce an impact assessment. In the meantime, critical systems may remain exposed or even knowingly left compromised to avoid service disruption.

---

[79] An incomplete list includes false-positive rate, false-negative rate, accuracy, precision, recall, mean-time-to-detection, and mean-time-to-remediation.

[80] A fundamental limitation in the evaluation of almost all detection technologies remains the determination of false-negative rates.

A related problem is determining the set of actions necessary to restore the integrity and trustworthiness of a system or network (along with the relevant data) after a compromise. To the extent that recovery is driven by damage assessment, there is an obvious dependency. Although one could theoretically imagine a fully agnostic system and network reconstitution (e.g., a full data and system recovery from a combination of full backup and reinstallation), several factors make this impractical at scale. These include critical external dependencies (e.g., credentials for external services), system and business availability constraints, and friction related to legacy or embedded devices (e.g., out of support devices) and failure of parts under stress (e.g., network saturation due to recovery traffic). Perhaps the biggest issue is the uncertainty in how far back to recover from,[81] especially as it pertains to data. As software systems become increasingly interdependent in both direct (e.g., cloud-enabled multi-device synchronization) and subtle ways (e.g., credential caching), the traditional fallback approach of reformatting and reinstalling becomes both untenable and insufficient.

The size and continuous piecemeal evolution of software, systems, and networks inhibits a sufficiently detailed understanding of their composition and functionality (even under attack-free conditions), which is a necessary step to identifying the aspects of system operation that were (or could be) accessed or tampered by an attacker. Combined with the inherent stealthiness of attacker activities, a timely reconstruction of a reasonably complete timeline of said activities and relevant system assets is currently infeasible except in limited situations. The telemetry or logging necessary to achieve the necessary degree of visibility, strongly correlated with but potentially more detailed than that needed for attack detection, can be cost- and performance-prohibitive to collect, store, and analyze in a timely fashion, even putting aside concerns about the integrity and reliability of the telemetry data in the presence of a sophisticated attacker. Furthermore, the need to restore or maintain system operations practically limits the time and resources that can be committed to the assessment analysis. In the (typically informal) risk analysis that drives the relevant parameters for the system recovery (i.e., which systems and data, how far back), this biases toward a focus on minimization of attack footprint, allowing for undiscovered latent access and other leave-behind artifacts introduced by the attacker. At the very least, it is important to capture in the final after-action report any specific assumptions made relative to the conclusions. For example, if the system was restored from a checkpoint created on a certain date, the inherent assumption is that the compromise occurred after that date.

In terms of potential solutions in the space of damage assessment, forensic reconstruction of attacker activities would benefit from the same type of solution as is needed for attack detection (see the section above on situational awareness), albeit with the

---

[81] Using an older backup is less likely to contain attacker artifacts, at the cost of lost data.

need for higher fidelity. Other relevant knowledge and capability gaps that need to be addressed include the following:

- Practical mechanisms are needed for creating high-integrity hot replicas or hot standbys that address common usage scenarios. Currently, the mechanisms and techniques that exist almost exclusively focus on server scenarios; expanding to desktop and mobile situations would greatly ease system reconstitution.
- Even if wholesale network recovery from scratch is impractical or infeasible, designing systems and architecting networks such that rapid reconstitution from scratch of key components and/or of large numbers of enterprise devices (e.g., by limiting or eliminating local storage and configuration) can help focus attention and resources to fewer locations. Relatedly, the development of techniques for incremental recovery is needed (e.g., prioritizing recovering and operating the parts of a system that are strictly necessary for business continuity).
- Determining and mitigating credential exposure, especially for those that are used by automated processes, is of high importance in reducing the potential for residual and lateral access. While credential management in general remains problematic, reconstitution in the absence of a human user presents its own particular challenges.

Several metrics can be used to gauge progress in this space. System-wide, the goal is to reduce mean time to second compromise and mean time to system restoration, and to achieve higher completeness in attack path reconstruction relative to ground truth (potentially in the context of red team–based evaluations, where ground truth can be made available).

This is a core aspect of cyber hard problem 10.

NEW TECHNOLOGY CYBER HARD PROBLEMS

New technology is bringing new cyber hard problems. A prime example is AI, another is CPS. The challenge in securing AI applications and CPS is a core contributor to cyber hard problem 9 but also affects 1, 2, 3, 4, and 10. It also profoundly affects cyber hard problem 7.

Security of Artificial Intelligence

It is difficult to assure the security of AI applications, particularly for design patterns that leverage generative AI. While the following discussion highlights AI-specific security challenges, it is important to note that AI systems are themselves software systems and thus susceptible to the full range of traditional cyberattacks.

The adoption of AI as an integral component in modern applications has been among the most disruptive innovations in computing this century. Many of the largest software companies have transitioned to using generative AI, as has become evident in public statements by Microsoft, Google, Meta, Salesforce, and others. Although traditional application security principles—when appropriately adopted—can safely accommodate the inclusion of AI components in software systems, there are unique attributes of AI that make securing forthcoming AI applications a hard problem.[82]

At the component level, both predictive and generative ML models are "non-smooth" systems that may produce very different outputs for similar inputs. Generative AI models are stochastic systems that can produce different inputs for the same input. Their non-smooth and sometimes stochastic nature may present a reliability challenge when using AI as a component in a repeatable system since their function cannot be formally guaranteed, nor behavior be fully characterized. Since the models themselves are not readily interpretable, this makes their safety and security difficult to assure. Remediation in AI components is difficult since the weaknesses which arise from training cannot be patched directly in code as it might in a traditional software component.

Applications using generative large language models (LLMs) typify several AI challenges. In a basic AI chatbot application, the user interacts with an LLM that iteratively predicts the next token (word chunk) from a growing input consisting of the original system instructions, user input, and previously predicted tokens. The initial and subsequent set of predictions is heavily influenced by the system instructions, which are designed to guide toward—but cannot robustly guarantee—predictions conforming to a preferred style or topic. Because LMs are instruction-following machines, attackers may attempt either indirectly or directly to lead the application away from the intended use. This can be especially problematic in agentic systems, in which the LLM output is connected to services that act on behalf of an (untrusted) user or respond to context fetched from external (untrusted) sources by the agent components.

While AI systems are fundamentally software systems, their characteristics—supply chains that include data sets and training code and runtime nondeterminism and non-smoothness—necessitate new approaches to risk assessment and vendor trust evaluation.

---

[82] A. Vassilev, A. Oprea, A. Fordyce, and H. Anderson, 2024, "Adversarial Machine Learning: A Taxonomy and Terminology of Attacks and Mitigations," NIST Computer Security Resource Center, January, https://csrc.nist.gov/pubs/ai/100/2/e2023/final.

To that end, many emerging regulations have begun to specifically call for "AI red teaming" requirements. Regulations in the European Union[83] and proposed regulations in the United Kingdom[84] legislate requirements for model assessment, with an emphasis on safety and societal harms. However, these requirements still lack acceptable standards across the industry in what should be assessed, what are acceptable assessment outcomes, and how and to whom to disseminate the results of an assessment.

Although there are fundamentally still software systems that include software and third-party services in applications, the supply chain of AI applications also includes data and third-party pre-trained or fine-tuned models. In addition to the possibility that attackers may develop model or data deserialization-based file formats (e.g., pytorch, pickle, and numpy) to execute arbitrary code,[85] the possibility exists that models may contain backdoor functionality encoded in the model's architecture or model weights. Technology and processes to measure and mitigate risk in these supply-chain components are nascent. Specific challenges in supply chain include (see more at NIST Adversarial ML Taxonomy[86]) the following:

- *Model assurance*—inherited vulnerabilities and weaknesses in third-party models, including the potential for deliberately backdoored models that rely on model weights rather than code for triggering mechanisms; and
- *Data assurance*—poisoning of open web–scale data sets used to pre-train or fine-tune models that can result in targeted or indiscriminate integrity violations.

A key challenge for auditability in the AI supply chain is that there is not yet a standard for reporting the equivalent of an SBOM, although efforts to address this have emerged.[87] The addition of model and data components can be accommodated by SBOM to include traditional static elements of component identification, dependency information, licensing, and versions. But for AI models, the behavioral reports should also be included that report on potentially risky runtime behaviors that have been

---

[83] European Commission, 2024, "AI Act," https://digital-strategy.ec.europa.eu/en/policies/regulatory-framework-ai.

[84] Department for Science, Innovation and Technology and the Office for Artificial Intelligence, 2023, "AI Regulation: A Pro-Innovation Approach," March 29, https://www.gov.uk/government/publications/ai-regulation-a-pro-innovation-approach.

[85] Common Weakness Enumeration, "CWE-52: Deserialization of Untrusted Data," https://cwe.mitre.org/data/definitions/502.html, accessed February 6, 2025.

[86] A. Vassilev, A. Oprea, A. Fordyce, and H. Anderson, 2024, "Adversarial Machine Learning: A Taxonomy and Terminology of Attacks and Mitigations," NIST AI 100-2 E2023, January, https://csrc.nist.gov/pubs/ai/100/2/e2023/final.

[87] J. Bressers, 2023, "SBOM Everywhere and the Security Tooling Working Group: Providing the Best Security Tools for Open Source Developers," Open Source Security Foundation (blog), June 30, https://openssf.org/blog/2023/06/30/sbom-everywhere-and-the-security-tooling-working-group-providing-the-best-security-tools-for-open-source-developers.

observed. Unfortunately, unlike a binary set of attributes or functions, the set of risky model behaviors may be incomplete, imprecise, and less actionable than in traditional software. Thus, it is still important to employ third-party audits and third-party guardrails to discover and control runtime behavior.

Defense Against Offensive Artificial Intelligence

Defenders are unprepared for a dramatic increase in scale and complexity of cyber operations from offensive AI tools—when attackers leverage AI for traditional cybersecurity operations. The risks presented by using AI for offensive purposes are offset at least to some degree by the potential for defenders to leverage AI to implement compensatory security controls and mitigations, but these are not addressed here.

As highlighted in the National Security Commission on Artificial Intelligence *Final Report*,[88] digital infrastructure may be increasingly indefensible against escalating, offensive, AI-enabled cyber capabilities without offsetting defensive controls. Threat actors are beginning to leverage AI for various malicious use cases, including offensive copilots, scaling social engineering attacks, and enhancing offensive operations.

Offensive AI is still nascent, but researchers are developing AI for various offensive purposes that will challenge defensive systems and processes. AI-driven offensive capabilities can increase the potency and speed of cyber campaigns and present significant threats to both digital infrastructure and human targets.

AI can expedite traditional cyber campaigns against digital infrastructure in several ways. For example, using LLMs, attackers can expedite the discovery, development, and delivery of exploits through automated code reversing, vulnerability discovery, and instrumentation of exploits for vulnerabilities. AI systems that reduce the time required for threat actors to execute attacks by automating labor-intensive tasks represent a sort of "offensive copilot" that can decrease the time to impact in cyber operations.

AI-powered tools can also assist attackers in more rapid maneuvering during hands-on parts of offensive campaigns to scale offensive operations. By integrating generative AI agentic frameworks with existing tools, attackers can orchestrate complex operations that cover large portions of an attack life cycle in a way that was not previously possible.

The impact on human targets using AI presents a formidable challenge. Disinformation campaigns that leverage deepfakes have already become part of public awareness due to several incidents involving elections[89] and digital warfare that now requires

---

[88] E. Schmidt, R. Work, S. Catz, E. Horvitz, S. Chien, A. Jassy, M. Clyburn, et al., 2021, *Final Report*, National Security Commission on Artificial Intelligence, released March 1, https://reports.nscai.gov/final-report.

[89] E. Sayegh, 2024, "The Battle for Truth in Election Seasons: AI-Generated Deepfakes," *Forbes*, May 14, https://www.forbes.com/sites/emilsayegh/2024/05/14/the-battle-for-truth-in-election-seasons-ai-generated-deepfakes.

news consumers to question the validity of reports.[90] Since bad news tends to travel faster than good news, correcting disinformation is an asymmetric challenge.

These tools can also be used for fraud. Highly realistic and interactive social engineering attacks for fraud are now possible with generative AI. In this setting, attackers can create convincing impersonations or scenarios to manipulate individuals in a way that feels customized.[91] AI's potential to scale such attacks is a developing threat vector, where generative AI can create deepfakes and other convincing forms of fake identities for automated and interactive phishing or scamming operations.

The human challenges that this presents have been called out in other cyber hard problems. The key ingredient that AI brings is the sophistication and potential for scale. While fundamental security practices can ward off many of these attacks, the increased scale and sophistication allows attackers and fraudsters to affect a much broader set of victims. In a setting of fixed resource constraints of defenders, remediation and response can become intractable.

Enforceable Policies for Data in Distributed Systems

Many cyber and cyber-enabled systems include a data component, either creating new data, processing existing data, or transmitting data to achieve a particular purpose. The designers of the system or application build the service with specific security and privacy properties to mitigate the occurrence and impact of adverse events—that is, uses of the data that go beyond the intended purpose. The user of the system, and relevant regulatory or law enforcement entities, desire the ability to hold the data steward accountable for upholding the properties as promised while also ensuring that unexpected uses of the data are not possible (i.e., "the software does what it says with the data; no more, no less").

In the absence of a solution to this problem, there is little choice but to trust that data are collected, used, and stored appropriately without much assurance. Prior to deployment or adoption, there is a requirement to convince the user and relevant authorities that the promised properties are sufficient and correctly implemented. Post-deployment, the data steward may need to modify the data use or protection terms and need to re-consent the data subject or owner, updating the presentation of the use and proposed protections and accurately recording the update.

---

[90] D. Klepper, 2023, "Fake Babies, Real Horror: Deepfakes from the Gaza War Increase Fear About AI's Power to Mislead," *Associated Press*, November 28, https://apnews.com/article/artificial-intelligence-hamas-israel-misinformation-ai-gaza-a1bb303b637ffbbb9cbc3aa1e000db47.

[91] H. Chen and K. Magramo, 2024, "Finance Worker Pays Out $25 Million After Video Call with Deepfake 'Chief Financial Officer,'" *CNN World*, February 4, https://www.cnn.com/2024/02/04/asia/deepfake-cfo-scam-hong-kong-intl-hnk/index.html.

Retrospectively, there is a requirement to be able to determine as much as possible about what went wrong (e.g., whether the security properties were inadequate, if the security properties were incorrectly implemented, or if the system were modified in some way that impacted the security properties).

- Example 1: A social network requests a phone number to be used only for MFA. How is the developer to build and prove that the phone number is used only for MFA and not targeted advertising or other contrary behavior?
- Example 2: The sustainability task force for a local municipality deploys a network of smart streetlights to reduce energy usage. The streetlight uses a video camera to determine lighting needs based on traffic patterns and natural light. How can citizens be sure that the local police have not used data collected by the cameras?
- Example 3: A personal genomics company sells a DNA testing kit and allows customers to opt-in to donating data to research trials. How can customers rest assured that the data are used only for the research trials they have opted into and that the results will not be subsequently used by health insurance companies?

A sizable portion of technical innovation is rooted in advancing the state of the art of what can be done with data, yet the technical mechanisms for setting and enforcing policies throughout the data life cycle (e.g., data at rest, in transit, and in use) have not kept pace. It is extremely time-consuming and difficult to identify, mitigate, and prevent the misuse of data without policy-aware data systems. In the past, we accepted possession and access to data as a proxy for permission to use the data. The proliferation of devices that collect data; the inherent complexity of the software, hardware, and network ecosystem; and also the ease with which data can be transmitted to another party make fine-grained control over the use of data untenable for the future. The misuse of data and our inability to make verifiable claims about how data will be used degrades trust in IT systems and hampers future innovations.

Making progress on this hard problem will support better outcomes on avoiding adverse outcomes for end users related to the misuse of data, increasing trustworthiness of personal devices, and perhaps decreasing disinformation. Large-scale change is needed to evolve IT systems to be policy-aware when processing data. At a minimum, such change requires the following:

- The design and implementation of a policy language that is suitably expressive (e.g., what is the agreed-upon set of "nouns" "verbs" and "attributes") and
- Designing policy authoring tools that meet the needs of diverse stakeholders, allowing them to write, review, negotiate, manage, and audit policies.

One open question is whether the desired policies are actually expressible (i.e., what kinds of policies are expressible and enforceable, and are these what people care about?). Although this has been tackled in traditional IAM systems (see above), it has not been done for shared data.

Technology-Enabled Disinformation and Fraud

The emergence of cyber-mediated, human-targeted attacks of various sorts has a history of being considered cybersecurity or privacy problems. Examples include inducing the receiver of a malicious email to download and run an attachment, click on a link (to deliver malware), clinking on a link and type things in (to steal identity), and the recognition of stalkerware as a category and problem. Targeting individuals through spear phishing and catfishing are recognized cybersecurity attacks.

Although propaganda, disinformation, and military deception have a long history, technology-enabled creation and dissemination of disinformation is a newer and growing problem. Everything on the computer is mediated by technology. The Internet and World Wide Web, along with social media, expands the reach of disinformation. Automation and AI expand the scale and precision of disinformation, to bots, deepfakes, and written and spoken text that can increasingly mimic anyone trustworthy. In parallel, "broadcast" sources of journalism (television, radio, newspapers) are being replaced with peer-to-peer communications with poor or missing authentication. What people see is determined in whole or in part by algorithms that (typically) optimize for engagement.

Per the Verizon 2024 *Data Breach Investigations Report*,[92] deepfake-like technology has already been used in many reported cases of fraud and misinformation. As generative AI only increases in abilities, scope, and scale, AI-generated fakes as weapons of disinformation will move beyond "deepfake" pictures of humans and audio fakes, to more complex scenes compellingly attesting to events that never occurred, compelling quotes, speeches, and "fake news" articles, and full videos. Generative AI that undermines artists of all professions today can become tomorrow's tools of disinformation.

Technology-enabled disinformation can be used to undermine individual reputations and emotional well being (e.g., deepfake revenge porn), create conspiracy theories

---

[92] Verizon, 2024, *2024 Data Breach Investigations Report*.

targeted at public figures, and attack core democratic processes such as elections. Much of the U.S. economy relies on the reputations of the strength of businesses and financial infrastructure, making this a potential weapon against U.S. economic stability.

What makes this a hard problem is the following:

- Core principles important to democracy, including free speech and privacy protections, preclude certain approaches to detecting these or fighting fire with fire. Censorship of U.S. citizens is directly counter to free speech.
- Notwithstanding the variously successful Internet censorship regimes, there are no geographical boundaries that serve as a natural defensible "perimeter."
- Misaligned incentives—for example, for ranking algorithms, "engagement" versus "truth." The former can be monetized. Unlike preventing cybersecurity attacks, preventing the use of a platform for disinformation is not unequivocally counter to the business concerns or agendas of those who own and run such a platform. Nor is it necessarily counter to the desires of those targeted (see *The Weekly World News* tabloid for a benign analogy and conspiracy theorists for a less benign example).
- Bots are often and usually indistinguishable from people, creating perceptions of scale and diversity that can be false and misleading.
- Human attributes and limitations are targeted: confirmation bias, the tendency to anthropomorphize, motivations, and desires to believe the best or worst in others.
- While there are experiments in the wild or with crowdsourcing and panels, there are no tools or frameworks to systematically consider how a deployed technology may be effectively used for disinformation, including threat modeling and abuse cases to consider the cybersecurity of a system.
- Using cyber means to maliciously manipulate individuals, communities, and societies for adversarial purposes is a cybersecurity problem, but not only a cybersecurity problem. It can involve social science, psychology, ethics, policy, business, law, and political science. It can overlap with other categories of potential cyber hard problems; social science of human capabilities and limitations and mitigations to account for challenges, engineering to mitigate human limitations and weaknesses, AI reliability, deepfakes, phones as a privacy problem, authentication and access control in the context of the global commons, and information and data provenance. It may involve so many disciplines and related issues that it is sui generis as a cybersecurity problem.

Information and Data Provenance

Content on the Internet (and later curated as part of another data set or AI function) often does not come with any trustworthy indication of the source or provenance of that information.

Much of the information people receive comes through someone else, either directly or through a communication artifact (e.g., books, newspapers). All the information received through computer interactions is the latter. Even a video call is intermediated by sophisticated software that can change backgrounds and faces. Information comes to consumers from or through a source, and their reaction to that information is potentially colored by knowledge of that source, from news to education to books, from religion to civics to politics. The reaction may be to the identity of the source, such as an individual (Walter Cronkite) or an organization (Fox News), or the reaction may be to the process and assumptions around the source's communication type (autobiography, medical advice).

The pseudonymity promised by "no one know[ing] you're a dog" is rapidly extending to all contexts on the Internet. "Fake news" is shared by people you know, and anyone can stand up a website claiming to be a news or publishing source. Identity, identity attributes, and source creation context are all at risk of being inaccurately relayed or assumed. Immersive virtual environments make alternate realities the norm.

Civil and societal institutions rely on some shared understanding of the authoritativeness of various kinds of information. Examples include news about communities, states, and nations, or results of the electoral process. The stability and safety of people's economic supports and investments rely on reliable information about them.

Building blocks for enabling a more trustworthy information ecosystem might include digital signatures (including source devices that apply signatures at the point of capture), imperceptible signals in media streams (watermarking), and widespread and reliable conveyance of provenance information through social media channels.[93] An alternate approach is centralized or decentralized fact checking and "community notes."

What makes this a hard problem is the following:

- Information flowing on today's Internet is almost always transformed in transit (everything from user-initiated cut and paste to transcoding to enable optimal use of bandwidth).
- Balancing strong provenance with privacy.
- News, as a money-making business, has almost disappeared, and there are now very few sources that are universally (or at least widely) trusted. News has also partly been replaced by entertainment. (Arguably, the news business

---

[93] Coalition for Content Provenance and Authenticity, "Overview," https://c2pa.org, accessed February 6, 2025.

was always funded by advertising, but now news organizations are not benefitting as much from the advertising revenue.)
- AI systems will be indistinguishable from humans for most online interactions and can scale to be formidable misinformation bots (e.g., AI systems summarizing source information). Weak authentication on the Internet, which makes it difficult or impossible to ascertain the source of a piece of information, exacerbates this problem.

POLICY CYBER HARD PROBLEMS

Many of the overarching cyber challenges described earlier are expressly amplified by missing or misaligned policies. It is difficult to design effective policy for complex systems that does not increase cost disproportionately to its benefit. Resilient system design and operations, even when properly guided by policy, can slow progress. Competing interests often retard policy solutions even when candidate solutions exist. Jurisdictional questions, including globalization and failures to provide federal preemption, further complicate policy solutions and effective remedies that would apply to providers and users in a single legal jurisdiction.[94]

Regulatory policy and economic incentives can be confounded by competing policy goals. For example, the desire for rapid problem identification can often be achieved by comprehensive authentication; however, this can often interfere with users' privacy. Furthermore, policy that encourages disclosure in support of principled risk assessment can threaten providers' intellectual property if done carelessly.

There has been policy progress that has helped ensure vulnerability and breach disclosure, but this is a fairly crude measure of resilience and safety.

The lack of effective policy (economic and regulatory) is one of the most dogged and influential of hard problems.

Policy hard problems need to be addressed by laws, policies, regulations. As described in the committee's overarching problem framework, they affect almost all cyber hard problems.

Misaligned Incentives

Misaligned incentives in cybersecurity are a significant challenge, manifesting in the varied and often conflicting priorities of stakeholders such as vendors, consumers, insurers, and regulators.

---

[94] Harmonizing policies across jurisdictions (state, federal, and international) is a super-hard problem.

Vendors prioritize speed to market and cost efficiency over security and say additional security measures would slow the pace of innovation, while consumers often choose products based on price rather than security features. Insurers, who have the potential to influence better security practices through underwriting conditions, struggle with accurately assessing risks and enforcing effective mitigations. Competitors, despite facing similar threats, are often unwilling to share valuable threat intelligence, undermining collective defense efforts. This misalignment results in suboptimal decisions that increase overall vulnerability and delay the benefits of addressing other cyber hard problems. Without progress on incentives, benefits from solving the other cyber hard problems will be disadvantaged or delayed.

Solving the issue of misaligned incentives is crucial for enhancing the overall security posture of the digital ecosystem. It matters to a wide array of stakeholders, including businesses that suffer financial losses from breaches, consumers whose personal information is compromised, and national and homeland security agencies tasked with protecting critical infrastructure. The economic impact of cyber incidents is substantial, with costs extending beyond immediate financial losses to include reputational damage, loss of consumer trust, and long-term recovery expenses. However, to date this has been inadequate to spur changes needed. Therefore, realigning incentives to promote better security practices is essential for reducing these risks and enhancing resilience against cyber threats.

The difficulty in addressing misaligned incentives stems from several factors. Economic and competitive pressures often discourage businesses from investing adequately in cybersecurity, as the benefits are not always immediately observable.[95] The tendency to prioritize short-term gains over long-term security investments is pervasive, and the lack of standardized metrics for measuring cybersecurity return on investment or cyber-coverage quality complicates decision making. Organizations may sometimes feel that the most cost-effective method for limiting damage for faulty products is public relations, especially for categories of weaknesses that are not readily assessed or repaired. Regulatory and policy efforts to realign incentives have been slow and fragmented, with various proposals such as grants, tax incentives, and liability considerations failing to achieve widespread implementation. The complexity of the cyber threat landscape and the rapid evolution of attack techniques further exacerbate these challenges.

Potential approaches to realigning incentives include policy reforms and innovative economic models. Governments can play a pivotal role by introducing and enforcing regulations that mandate minimum security standards, introducing "safe havens"

---

[95] This is related to *myopic loss aversion*. See R.H. Thaler, A. Tversky, D. Kahneman, and A. Schwartz, 1997, "The Effect of Myopia and Loss Aversion on Risk Taking: An Experimental Test," *The Quarterly Journal of Economics* 112(2):647–661.

for good faith efforts accompanied by product design transparency, and by offering tax incentives or subsidies for businesses that invest in robust cybersecurity measures. Public–private partnerships can facilitate better information sharing and collective defense initiatives. An example of a public–private partnership that does such a thing successfully is the National Cyber-Forensics and Training Alliance, which brings together the business sector and law enforcement to disrupt cybercrime. Additionally, developing standardized metrics for assessing cybersecurity investments and outcomes can help businesses make more informed decisions. The implementation of mechanisms like the U.S. Cyber Trust Mark, which provides a recognizable standard of cybersecurity for consumers of wireless IoT devices, is a step in the right direction.

Those who can take action to realign incentives span across sectors. Policy makers and regulators can introduce and enforce laws that require graded security standards, depending on the kind of device and its use environment, and incentivize compliance. For example, autos that can be easily stolen because their keyfobs use weak cryptography and pervasive back doors in network-connected devices seem like areas that need to be addressed. Absent this, industry leaders and business executives will not prioritize cybersecurity as a critical component of their operational strategy and allocate appropriate resources. Insurers can refine their risk assessment models and offer premium reductions for policyholders that adopt best practices. Consumers can influence the market by demanding more secure products and services. Additionally, cybersecurity researchers and advocacy groups can continue to highlight the importance of aligned incentives and drive awareness.

Success in realigning incentives can be measured through several indicators. A notable decrease in the frequency and severity of cyber incidents would suggest that stakeholders are making more security-conscious decisions. Increased investment in cybersecurity by businesses, moving closer to the recommended 10 percent of budgets, would also be a positive sign. Ultimately, success will be reflected in a more resilient and secure digital ecosystem where the costs and benefits of cybersecurity investments are better aligned across all stakeholders.

Cybersecurity Poverty

An often-overlooked consequence of technology's spread is the difficulty that organizations and individuals have in securing it. Originally described in 2011,[96] the "security poverty line" is a concept that delineates the "haves" from the "have nots": whether it is economically or technically feasible to implement what is generally assumed to be effective security, given real-world conditions. Just as with economic poverty, cybersecurity

---

[96] W. Nather, 2011, "T1R Insight: Living Below the Security Poverty Line," 451 Research, May 26, https://web.archive.org/web/20140203193523/https:/451research.com/t1r-insight-living-below-the-security-poverty-line.

poverty results from many complex dynamics and factors. This problem exacerbates the effects of cyber hard problems 1, 2, and 4.

Because there is no simple prescriptive blueprint for building secure systems, some have tried to measure effective mitigation of carefully scoped attack scenarios, such as MITRE's Engenuity evaluations[97]; others have tried to calculate the projected cost of security technology according to a given compliance framework[98] or simply following security professionals' recommendations.[99] Although peer benchmarking and trends reports describe how much some organizations spend on cybersecurity, the reports do not address whether the spending is effective or appropriate. Spending formulas, such as the percentage of IT budget, do not necessarily scale up or down, nor do they have any link to positive or negative outcomes. Some increasingly critical controls (e.g., logging) are not included in the minimum baseline edition of products but are premium priced.

Another confounding factor for organizations is expertise. Cybersecurity expertise is not simply education or training; it also includes the experience of securing new technology and diagnosing and responding to new vulnerabilities and attacks. Organizations find themselves competing for this expertise against the deeper pockets of security providers (according to Glassdoor, the total salary in 2024 for a senior cybersecurity analyst is $156,000–$234,000 per year).

Constraints within the environment also affect an organization's capability to secure itself. For example, conventional best practice in cybersecurity calls for a system to be designed to fail safe rather than open; this is not an option in a safety-focused sector such as health care, where medical staff may never be barred access to equipment or data needed to treat patients in an emergency. Software that integrates with hundreds of different systems under a variety of countries' regulatory environments can take months or years to update. Onsite upgrades for thousands of point-of-sale systems mean that retailers must choose carefully when to incur that downtime and expense, and certainly not during the heaviest shopping times of the year. Every cybersecurity risk framework or practice may need to be adapted substantially to work around these obstacles.

Finally, in an era where cybersecurity controls are spread among third-party providers (see the section "Supply Chain" below), organizations have to rely on the cooperation of other entities with whom they may have little to no legal or commercial influence. With a sufficiently large amount of money at stake or the possibility of negative public relations, a provider may be incentivized to meet the security requirements of a customer, but smaller organizations lacking this kind of influence cannot necessarily receive

---

[97] MITRE, 2024, "Our ATT&CK Evaluations Methodology," https://attackevals.mitre-engenuity.org.
[98] Center for Internet Security (CIS), 2023, "The Cost of Cyber Defense," CIS Controls Implementation Group 1, August, https://www.cisecurity.org/insights/white-papers/the-cost-of-cyber-defense-cis-controls-ig1.
[99] A. Shimel, 2013, "What Is the Real Cost of Security?" NetworkWorld, April 4, https://www.networkworld.com/article/744780/opensource-subnet-what-is-the-real-cost-of-security.html.

the emergency services they need during an incident, force the timely remediation of an identified vulnerability, or reject provider conditions that may result in increased risk (such as allowing overly broad network access). Regulations will not work unless the regulated parties have access to tools that are affordable (sustainable) and actually reduce risk. Without uniform cybersecurity regulations or other incentives, most small- and medium-sized businesses, nonprofits, and local public-sector entities (including law enforcement) have to make do with the equivalent of security scraps, with support only available piecemeal from managed service providers, from a provider specified by their cyber insurer, or volunteer efforts such as the University of California, Berkeley–led Cybersecurity Clinics.[100]

Supply Chain

Fully organic development shops such as Google and Apple have the advantage of full (internal) transparency in their software code bases. This facilitates direct analysis at scales ranging from lines of code in small components to design choices for APIs and architectural features. This also facilitates comprehensive assured refactoring; for example, updating a service API with potentially hundreds of clients, all incompatibly updated in an atomic action. Additionally, it facilitates a fully explicit linking of design models, implementation artifacts, test cases, analysis tooling, and any supporting elements.[101,102]

In other words, full transparency facilitates ongoing acceptance evaluation, rapid adaptation, and repairs without creating technical debt (i.e., expedient decisions that would later need to be revised in order to permit continued evolution of a system).

By contrast, large enterprise and mission systems are generally integrated from diversely sourced components and services ("system elements"), some of which are kept opaque to their clients in order to retain competitive advantage, protect sensitive data and algorithms, and enable update and enhancement without unwanted dependencies on (hidden) implementation choices. This means that even when one layer is revealed in a complex system, there can be multiple opaque layers beneath, analogous to "turtles all the way down."

The integrated systems model poses challenges, however. One set of challenges relates to acceptance evaluation, due to opacity of system elements and uncertainty regarding compatibility of elements. Another set of challenges relates to update and evolution, deriving from compatibility issues as individual elements are on uncorrelated update cycles. (Services, for example, can be updated several times per day, while

---

[100] Consortium of Cybersecurity Clinics, "Cybersecurity for the Public Good," Center for Long-Term Cybersecurity, https://cltc.berkeley.edu/program/consortium-of-cybersecurity-clinics, accessed February 6, 2025.
[101] "Why Google Stores Billions of Lines of Code in a Single Repository," posted September 14, 2025, by @scale, YouTube, https://www.youtube.com/watch?v=W71BTkUbdqE.
[102] H. Wright, 2019, "Lessons Learned from Large-Scale Refactoring," *2019 IEEE International Conference on Software Maintenance and Evolution*, December 5, https://ieeexplore.ieee.org/document/8919159.

open-source components may be updated every few weeks.) Evolution is also impaired by opacity, and many systems sustainment teams must engage in active reverse engineering to assess and document for repair security vulnerabilities, for example. There are also challenges related to overall systems architecting and design. Architectural decisions focused on reducing coupling, localizing variabilities, enhancing key quality attributes (particularly resiliency), and the like may need to be compromised to support compatibility of APIs, data representations, and service interfaces among elements that are meant to interoperate.

An extreme example is the incorporation of vendor components as original equipment manufacturer elements into integrated systems, such as commercial desktop systems into medical devices such as imaging systems. The end user, and possibly the IT support team, might not have sufficient visibility to be aware of the incorporated desktop as other than part of an appliance, and so that desktop may not, over a period of years, receive necessary updates and security patches. The resulting vulnerabilities have been exploited in ransomware attacks.[103,104]

An additional consequence of supply chain opacity is hidden dependencies, where a deeply embedded vulnerable system element can trigger disruptions in the event of compromise or, in the case of open source, loss of configuration control. Attacks on embedded supply chain elements can have broad consequences, and so these elements are a favored target by attackers. Examples include Blackbaud, a service provider to financial services and other organizations including critical nonprofits.[105] A research report by the Cyentia Institute (a subsidiary of Mastercard) and RiskRecon[106] identified ripple effects impacting between 800 and 1,000 downstream organizations. The network security vendor SolarWinds unintentionally delivered a malware payload embedded in a signed system update that was automatically distributed.[107] A more recent extended global outage, caused by an automatically deployed update to CrowdStrike security software on Windows systems, affected millions of systems from banks to commercial aviation, health care, and critical infrastructure.

This can be an issue even when the embedded system element is a tiny fragment of code. One example from 2016 is leftpad, which is an 11-line module of code in the

---

[103] L. Hautala, 2020, "Hospital Devices Exposed to Hacking with Unsupported Operating Systems," *CNET*, March 10, https://www.cnet.com/health/medical/hospital-devices-exposed-to-hacking-with-unsupported-operating-systems.

[104] C. Van Alstin, 2023, "RSNA 2023: Hospital Imaging Systems May Be Gateways for Ransomware, Expert Warns," *HealthImaging*, November 30, https://healthimaging.com/topics/professional-associations/radiology-associations/radiological-society-north-america-rsna/rsna-2023-ransomware-medical-devices.

[105] L. Fair, 2024, "FTC Says Blackbaud's Lax Security Allowed Hacker to Steal Sensitive Data—and That's Just the Beginning," Federal Trade Commission (blog), February 1, https://www.ftc.gov/business-guidance/blog/2024/01/ftc-says-blackbauds-lax-security-allowed-hacker-steal-sensitive-data-thats-just-beginning-story.

[106] Riskrecon, "New Report: Ripples Across the Risk Surface," Riskrecon by Mastercard, https://www.riskrecon.com/ripples-across-the-risk-surface, accessed February 6, 2025.

[107] L. Fair, 2024, "FTC Says Blackbaud's Lax Security Allowed Hacker to Steal Sensitive Data."

million-element open source NPM ecosystem widely used for web applications. The developer of this small component chose to delete it and other elements from the library due to a dispute over names for a software package. The deletion lasted only 2 hours but caused widespread disruption because of its pervasive use deep in the supply chain supplying web applications.[108]

There are also supply-chain issues in IoT, CPS, and networking infrastructure generally. One example was an attack on small office and home office (SOHO) routers, identified by Black Lotus Labs at Lumen Technology.[109] In this case, more than 600,000 routers belonging to a single internet service provider were completely disabled, forcing the entire customer base to have their equipment physically replaced. In other cases, these SOHO routers are not regularly updated or are no longer supported by the vendor so that no security updates are available; the accumulating residue of vulnerabilities makes a perfect platform for attackers to take over infrastructure and use it for botnets or proxying services.

On the one hand, the approach to security vulnerabilities has been to encourage organizations to patch early and often, and preferably automatically. But as these examples show, automatic updates gone wrong can also cause catastrophic events. Victims of attacks and outages are caught in the middle between conflicting imperatives, and the cybersecurity industry owes them a better answer than to say "just patch."

The recent mandates regarding use of an SBOM can be seen as transforming what is sometimes a full opacity into a kind of "translucency" where some information is provided downstream (i.e., to client users, integrators, and end customers) in order to overcome some of these challenges and, additionally, create some incentives within the supply chain to address security attributes more aggressively. An SBOM, representing something akin to a food ingredient list, can empower organizations to make better procurement decisions, but only when they have feasible alternatives.

This is, of course, a supply-chain cyber hard problem also affecting hard problems 1, 2, and 3.

Liability

Liability for faulty code or hardware represents a critical and complex issue in cybersecurity. Under "contracts of adhesion" vendors often sell software "as is," disclaiming responsibility for defects that may lead even to significant breaches or failures. Cyber systems cannot be evaluated based on a quick inspection (like a vacuum cleaner) or even a diligent inspection. Choices for equivalent functionality in other products, are often

---

[108] Ibid.
[109] Black Lotus Labs, 2024, "The Pumpkin Eclipse," Lumen, May 30, https://blog.lumen.com/the-pumpkin-eclipse.

very limited and, as discussed, there is a market failure that does not practically enable consumers to "select the model with the security they want." Except for copyright and patent infringement and designated systems like medical devices and automotive, or software used in other regulated industries, where limitation of liability is circumscribed by law, most software is marketed with the "understanding" that some level of imperfection is acceptable. Thus "legal" remedies are largely ineffective even with expensive litigation. This becomes particularly problematic in cybersecurity where software faces sophisticated, evolving attackers. The challenge is compounded by the displacement of loss onto consumers rather than the companies responsible for the vulnerabilities.

The consequences of faulty code extend far beyond mere inconvenience. Consumers, businesses, and governments all suffer from the fallout of software failures. Establishing liability would incentivize vendors to prioritize security and quality, potentially reducing the frequency and severity of breaches. However, current practices and economic realities pose significant hurdles. Large companies, even after significant breaches, rarely face existential threats, and the costs are often borne by consumers and smaller entities.

Addressing this issue requires overcoming several barriers. First, there is a need for a cultural and operational shift within the software industry. The mantra that "we did the best we could" (even if true) must give way to more rigorous standards and accountability. Introducing liability necessitates robust metrics and frameworks to evaluate software safety and security, akin to those in place for other regulated industries. However, creating these standards is not straightforward. The dynamic nature of software development, coupled with the continuous evolution of threats, makes it difficult to establish a static baseline for safety. Furthermore, the global nature of the software supply chain complicates the assignment of liability, as many stakeholders—from developers to suppliers—are involved in the creation and maintenance of software products.

Potential approaches to this problem include both voluntary and mandatory assessment mechanisms. These standards would need to be continuously reviewed and updated to remain relevant. Another approach is to create a "safe harbor" for vendors who follow best practices, thus incentivizing compliance while recognizing the inherent challenges of achieving absolute security.[110]

Government agencies can establish and enforce regulatory standards, while industry groups can develop and promote good practices, with incremental adoption. Companies, particularly large enterprises with significant market influence, can lead by example, incorporating security into their development processes and advocating for broader industry changes. Collaboration between public and private sectors is essential

---

[110] Lawfare (https://www.lawfaremedia.org/topics/cybersecurity-tech) has a number of relevant notes on this topic.

to ensure that standards are practical and effective. Although other domains such as medicine and civil and mechanical engineering have successfully employed professionalization standards, the software field generally does not possess the sort of widely accepted, comprehensive principles as the other fields.[111]

Success in this endeavor would be indicated by a measurable reduction in the frequency and impact of software-related breaches. Metrics could include the number and severity of vulnerabilities discovered and exploited, time to patch after exploit announcements, the financial and operational damage from breaches, and the rates of compliance under established standards. Ultimately, creating a more secure software ecosystem will require sustained effort and cooperation across the industry, but the potential benefits for all stakeholders make it a goal worth pursuing.

Third-Party Intervention

Most cyberattacks rely on communication over the Internet. The Internet, globally and even within countries, is not owned, operated, or controlled by a single legal or technical entity or by a closed set of governments, major corporations, or technology institutions; nor is it subject to a single set of policies. Rather, the Internet is an aggregate whose owners, operators, participants, and technologies function independently of one another but interoperate. As a result, long before they touch their target networks and endpoints, cyber operations traverse and in some cases leverage infrastructure, technology, and services owned, operated, and offered by different Internet infrastructure providers (IIPs). These companies include (without counting computing hardware companies) operating system developers, cybersecurity firms, Internet service providers, mobile telecommunications companies, cloud and virtual private server providers, content delivery networks, Domain Name System service providers, hosting providers, domain registrars, and a variety of Internet technology platforms, such as browser, e-mail, and search platforms.

These companies see themselves as neutral providers of global Internet services but not as critical infrastructure assets with a key role to play in systemic security and resilience. Although most of these companies invest substantially in cybersecurity measures, and occasionally cooperate operationally to degrade specific threats, they lack regulatory, financial, or other incentives to systematically address malicious actors' use of their technology and services in cyber operations that do not directly and immediately impact them. Their cybersecurity efforts and cross-company cooperation are further hamstrung by a complex legal, privacy, regulatory, antitrust, government, and business environment.

---

[111] Circumscribed certifications, like network administration, can be useful but they are scarcely comprehensive.

Internet infrastructure companies have unique (and in many cases the only legal) vantage points to observe attacks and detect and frustrate malicious cyber activity on a systematic, nationwide basis (and, for some of the large IIPs, even globally). If these entities were to take on the responsibility of discovering and limiting malicious cyber activities and implement successful regimes to do so, malicious actors' ability to conduct cyber operations against U.S. targets would be substantially attenuated. Coordination across providers is required because attackers straddle their infrastructure across multiple providers for survivability against takedowns.

There have been several attempts at private-sector coordination (with or without U.S. government participation), primarily at the threat information sharing (TIS) level, over many years. *Perhaps the biggest inhibitors to such coordination and threat information sharing revolve around liability concerns and the lack of a business case.* While setting industry-wide standards and the exchange of best practices is non-controversial, the (at best) federated nature of the infrastructure landscape means that operational coordination that leads to systemic action (e.g., coordinated takedowns of malicious infrastructure) is the exception rather than the norm, and only occurs as a knee-jerk reaction to high-visibility events. Therefore, in practice, most coordination has taken the form of TIS. This need not be a handicap, if the information exchanged is accurate, timely, and conveys sufficient context to provide the necessary confidence so each participant can take action. Unfortunately, that is typically not the case, and critical information is often missing due to lack of collaboration across key industry verticals.

One of the potential liability concerns expressed is that TIS represents or can lead to collusion between companies (creating the perception of collusion to the public and to regulators), leading to privacy and antitrust concerns (and associated litigation). Making the government part of any threat information sharing arrangement focused solely on technical factors is one possible approach, but for the global firms, there will also be concerns on other countries' reactions to such direct U.S. government participation. The most common liability concern comes from actions following high-profile attacks. All such incidents have been accompanied with various types of litigation, including litigation by different agencies of the U.S. government. For companies, the most damaging form of litigation alleges "willful neglect," meaning the commercial entity knew or should have known about the specific problem but failed to act. This creates a perverse incentive for a deliberate lack of knowledge in some area as a cost-effective form of liability protection. There will be no long-term continuous incentive for TIS or any other meaningful collaboration without addressing these concerns.

There are at least four different concerns over liability: customer and third-party impacts; notification shortfalls; regulatory fines; and civil and class-action lawsuits. In many cases, this involves different agencies such as the Department of Justice, the

Securities and Exchange Commission, the Federal Trade Commission, the Department of Homeland Security Cybersecurity and Infrastructure Security Agency (CISA); and, in certain cases, the Department of Defense. While CISA has made progress with information sharing, the fear of liability still exists because it is not possible for CISA to fully protect companies reporting a breach from litigious actions taken by other government agencies due to different authorities and regulations. In addition, CISA cannot protect companies from civil lawsuits from other companies, customers, and third parties.

Accordingly, and starting from a recognition that its greatest strength is as a convener for collaboration on a technical level, the U.S. government could provide a safe, non-litigious forum for technical collaboration and data sharing without fear of liability from both the U.S. government and private industry, perhaps after the model of the Information Sharing and Analysis Centers.[112] Such a forum would create potential for end-to-end visibility across the domestic Internet infrastructure. Such a forum could be coordinated with the U.S. Cyber Command, and potentially be tipped by the Intelligence Community—but, critically, it should not be used as a source for data by the latter. To the extent that technologies for private TIS exist or can be developed, they will play a significant role in countering narratives of collusion and negative public perception.

One option is to require all companies over a certain size to have cybersecurity insurance—and allow companies to work together to bundle end-to-end coverage to further sharing of information. This also creates the incentive for companies to follow standards and adopt technologies for lower insurance rates. Working through the insurance providers may be a more tractable proposition. However, cyber insurance coverage is very limited and for all the reasons mentioned earlier, insurance companies themselves are in no position to judge the security of the systems they insure. Careful mechanism and incentive design are needed to avoid simply transferring the intractable problem from the company to an insurance firm.

Individuals and organizations have a legal right to pursue those who violate their service agreements with civil courts. Creating a special cyber court and providing it with the necessary technical resources to fully pursue cyber criminals and threat actors (even nation-state sponsored ones, to the extent that it is not desirable to treat these as an act of war) may be appropriate given the technical understanding required by the U.S. government and all parties.

Another option is to create a "cyber fire department," with broad authority to act on third-party, including private cyberinfrastructure. This could be operated by the U.S. government, by contracted private entities, or in a federated or localized manner to reflect constraints of specific sectors, geographical areas, or other considerations. It is

---

[112] National Council of ISACs, "Information Sharing and Analysis Centers (ISACs)," https://www.nationalisacs.org, accessed February 6, 2025.

almost certain that significant new authorities conferred by legislation (including limited liability waiver) would have to be granted to such an entity.

Ultimately, if the United States wishes for the Internet infrastructure providers to play better defense, it will have to either create the right financial incentives for markets to value security more than they currently do, or directly pay for such better practices. Tax credits, bounties, fees on connectivity bills, security investment programs through the Small Business Administration, and subsidized cyber insurance (combined with heightened terms and conditions for such policies) are only some of the ways such financial support can be extended.

5

Toward Community Coordination and Progress

Cybersecurity is a fundamental societal problem with extraordinary scope and dimensionality, including technical, human, business, and policy. The challenge is exacerbated by the difficulties of measurement. Without good ways to measure security, incentives are hard to frame, and so action becomes difficult to motivate, especially when there are benefits to expediency. For some of the previously identified cyber hard problems, there has been extensive technical progress. These levers are important and essential, and indeed there are strong reasons to expand and accelerate technical work. But transformational impact on many of the hard problems identified in this report requires, additionally, collaboration with stakeholders involved with business and policy.

Driving this is that security is an attribute of a cyber system within a context of operation. A cyber system includes hardware, software, services, and associated supply chains. It also includes how a system and its associated data interact with other systems. The context of operation includes the ways in which the system interacts with the world, including with human users and operators, as well as with other systems—and, importantly, with potential cyber adversaries who may seek access to confidential data, or ability to tamper with data, or disruption of operations. Examples of context of operation range from civil infrastructure operations and national security systems to consumer applications for financial services, health care, and online media.

UNDERSTANDING AND MEASURING PROGRESS

The previous chapters have described pervasive difficulties in measuring security attributes, challenges that affect the ability to frame incentives for stakeholders, especially when counter incentives are present, such as cost and schedule (which also have the benefit of being easily measurable). Thus, making progress on assessments and related interventions is essential and valuable. Here are some elements of this broad challenge of framing and defining measurable security-related attributes.

1. *Cyber risk.* Prioritizing security interventions requires a comprehensive look at the three principal elements of cyber risk:
    a. Reducing vulnerability of the system itself through engineering choices and using modeling and analysis to support assessments. There are many kinds of vulnerability, and each has associated models, analysis techniques, and preventive engineering practices.
    b. Reducing potential consequence of compromise through adjustments to the ways systems can interact in the world, including the ways they interact with people.
    c. Understanding potential threats that are associated with the role of that system and society and with the attack surfaces that are exposed (as a consequence of choices regarding both system design and context of operation). Security is distinguished from most engineering disciplines because it must always anticipate a well-resourced, capable adversary motivated to undermine security for some benefit. When vulnerabilities in a system cannot readily be repaired, a natural step is to alter the operational role of that system to reduce exposure of attack surfaces and consequences of compromise.
2. *Transparency in support of evaluation.* Security is a complex aspect of an entire system. In an evaluation process, information asymmetry (the differences between what a producer knows and what a consumer knows) creates challenges, especially when powerful producers have rational incentives both to thwart transparency and to benefit from this to game perceptions to their benefit. Even with full transparency, these aspects of risk are difficult to assess and improve and will likely remain difficult.
    a. Direct evaluation of opaque "black box" systems though testing will not yield reliable results with regard to security.

b. Direct evaluation with transparency—supported by models and evidence—is the only way to achieve fully confident judgments and reasonable trade-offs regarding specific aspects of security.

c. Indirect evaluation based on engineering processes rather than on engineered artifacts will also not yield reliable results. However, evidence of attention—in systems design and in scoping of operational context—to security engineering principles, such as least privilege, can increase confidence.

In many areas, barriers of scale and usability are being successfully challenged to advance our capacity for direct and fully trustworthy judgments about particular security-related quality. These steps apply regardless of whether it is a national security system, a social media service, or a consumer-facing mobile app.

A particular challenge in both evaluation and operation is the capacity of sophisticated adversaries to gain access to systems and their supply chains in ways that may often afford them greater knowledge of the details of a system and its elements than operators, evaluators, and sometimes even developers (when there is opacity in the supply chain). Highly capable adversaries may synthesize both a more holistic and a more detailed knowledge about a system than its users, operators, evaluators, and even developers.[1]

3. *Increments of transparency*. The challenge of security assessment is exacerbated by adverse business, legal, and regulatory norms. In the absence of security measures, business incentives (cost, schedule, minimized liability) lead to insecure systems. Small improvements in measures and in transparency, for example, the recent efforts to encourage a software bill of materials and "zero trust" practices, can result in both security improvements and in momentum to further transparency. There are a few security attributes that can be reliably assessed, if attested to by a third party. Experimentations are to be encouraged in this regard. Rewarding transparency and evidence-based attestations in acquisitions will likely be helpful.

4. *Targeted compliance standards*. Generic security standards and frameworks are not necessarily straightforward to implement in various constrained environments. Although some verticals have created specific compliance standards for regulatory purposes, there's a need for more extensive development of "templates" and mitigations that have been shown to work under various technical and economic constraints. The MITRE organization has been doing

---

[1] Cybersecurity and Infrastructure Security Agency, 2025, "Closing the Software Understanding Gap," January 16, https://www.cisa.gov/resources-tools/resources/closing-software-understanding-gap.

strong work in this area, but it needs to include adaptation for small and medium businesses and other resource-poor entities.

5. *Assessing increments of progress.* Security challenges are rarely "solved," but rather are subject to increments of progress. Password composition practices provide an example. For many years, there was a widely held belief that complex passwords, frequently changed, were the best practice.[2] For users without password managers, this meant a compromise on usability, since these complex passwords (multiple special characters, numerals, upper and lower case, etc.) were difficult to remember and to type. Indeed, there was folk wisdom that, regarding user authentication, one could have either security or usability, but not both. Empirical studies subsequently demonstrated that longer passphrases could be both memorable and secure.[3,4] This disrupted the folk wisdom and gave way to improved National Institute of Standards and Technology password guidance.

It is daunting that the cyber hard problems are complex and do not lend themselves to simple "magic bullet" solutions. Some say that cybersecurity is a marathon and not a sprint, but for many of the problems, the struggle will be ceaseless—as systems grow in complexity and ambition, and cyber adversaries grow in capability and motivation. Choices will have to be made about managing a portfolio of research to make progress on multiple fronts. There is no choice but to engage in the struggle, and with relentless vigor—since the capabilities gained from cyber systems are now essential to advancing health, education, communications, commerce, national security, scientific research, and information dissemination. Cyber systems enhance productivity and create new kinds of economic value—and there is no limit in sight in the scope and impact of what they can be doing in the future.

---

[2] National Institute of Standards and Technology, 2024, "SP-800-63B—Authentication and Lifecycle Management," *Digital Identity Guidelines*, August 28, https://pages.nist.gov/800-63-4/sp800-63b.html.

[3] S. Komanduri, R. Shay, P.G. Kelley, M.L Mazurek, L. Bauer, N. Christin, L.F. Cranor, and S. Egelman, 2011, "Of Passwords and People: Measuring the Effect of Password-Composition Policies," *Proceedings of the SIGCHI Conference on Human Factors in Computing Systems* 2595–2604, https://www.ece.cmu.edu/~lbauer/papers/2011/chi2011-passwords.pdf.

[4] P.G. Kelley, S. Komanduri, M.L. Mazurek, R. Shay, T. Vidas, L. Bauer, N. Christin, L.F. Cranor, and J. Lopez, 2012, "Guess Again (and Again and Again): Measuring Password Strength by Simulating Password-Cracking Algorithms," *2012 IEEE Symposium on Security and Privacy* 523–537, http://ieeexplore.ieee.org/iel5/6233637/6234400/06234434.pdf.

INFORMING RESEARCH INVESTMENTS AND POLICY ACTIONS

The cyber hard problems identified in this report can be used by various sectors and entities as reference for research and development (R&D) investments as well as incentives and policy actions.

- *Academic researchers* in cybersecurity can frame their work in the context of the hard problems to connect with partners working on other aspects—technical, policy, business, societal, and so on. This is because, as noted, many aspects of the problems—and potential progress toward solutions—are not purely theoretical or technical. Researchers engaged in multidisciplinary studies (i.e., cybersecurity plus law, policy, international relations, engineering, psychology, economics) can benefit both by helping in framing the overarching problems and also providing applicable wisdom from their disciplines.
- *Industry* is at the forefront of both the creation and the use of the great diversity of societally relevant cyber systems. On the demand side, there are challenges in evaluation and acceptance evaluation, defining operational workflows, and integrating systems into a broader enterprise including other systems. On the producer side, there are challenges both in addressing the technical requirements and appropriating technical advantages while also providing customers and users with evidence to support their confident acceptance of the products and services offered. This is important both for established players and for new entrants. Additionally, these cyber hard problems provide a framing to facilitate targeted R&D collaboration with higher education and for sector-wide initiatives.
- *Policy makers and federal research funding agencies* can look at the cyber hard problems list as a framing for sponsorship of strategically relevant cyber research, in both the mission-focused mode and the exploratory mode. In the mission-focused mode, many of these areas directly affect national security—and not just in the sense that cybersecurity failures create cascading effects across critical infrastructure and economic activity but also in the enactment of the continuous and often intense engagement with cyber adversaries. Opportunities exist for incentives and regulation to address cybersecurity risk that now affects nearly all sectors and consumers. The cyber hard problems also can provide useful framing for international alignment of cybersecurity policy and processes for resilience. This can complement the de facto alignment that is a consequence of the global presence of tech firms.

Appendixes

A

Statement of Task

A National Academies of Sciences, Engineering, and Medicine consensus study will identify key "hard problems" for cyber resiliency. The study will build on prior work sponsored by the Department of Homeland Security—a "Cyber Hard Problem List" first issued in 1995 by the now-sunsetted federal InfoSec Research Council and most recently updated in 2005.

The committee's report will:

- Provide a current list of hard problems in cyber resiliency, building on earlier hard problems lists and expanding the scope from cybersecurity to cyber resilience;
- Describe the criteria and process used to select the hard problems;
- Assess impact and evolution of earlier hard problem lists including influence on cybersecurity research and development (R&D) investments and activities, progress toward solutions, and items added or removed over time; and
- Identify ways that the new list could be used to enhance community-wide coordination of R&D activities.

In addition to the consensus study report, a short dissemination product will be prepared that summarizes the list of hard problems.

B

Briefings to the Committee

The committee received briefings from the following presenters:

Thomas Berson (NAE), Salesforce
L. Jean Camp, Indiana University
Tony Cheesebrough, Cybersecurity and Infrastructure Security Agency
Jim Dempsey, University of California, Berkeley, School of Law
Drenan Dudley, Office of the National Cyber Director
Kathleen Fisher, Defense Advanced Research Projects Agency
Tom Forest, General Motors
Alex Gantman, Qualcomm Technologies, Inc.
Lawrence Gordon, University of Maryland
Eric Grosse, Independent Consultant
Jason Healey, Columbia University
Cormac Herley, Microsoft Research
Chris Inglis, U.S. Naval Academy
Lance Joneckis, Idaho National Laboratory
Paul Kocher (NAE), Independent Researcher
Carl Landwehr, University of Michigan
Ian Levy, Amazon
Steve Lipner (NAE), SAFEcode
Olga Livingston, Cybersecurity and Infrastructure Security Agency
Martin Loeb, University of Maryland
Mark Montgomery, Cyber Solarium Commission

Fred Schneider (NAE), Cornell University

Nick Selby, Evertas

Adam Shostack, University of Washington

Phil Venables, Google

John Viega, Crash Override

Paul Waller, National Cyber Security Centre

C

Committee Member Biographical Information

JOHN MANFERDELLI, *Chair*, is a principal at Datica Research. Before that, he was the confidential computing Incubation project leader in the Office of the chief technology officer (CTO) at VMware. Prior to VMware, he was a professor of the practice and the executive director of the Cybersecurity and Privacy Institute at Northeastern University. Immediately prior, Manferdelli was the engineering director for production security development at Google. Prior to Google, he was a senior principal engineer at Intel Corporation and the co-principal investigator (PI) (with David Wagner) for the Intel Science and Technology Center for Secure Computing at the University of California, Berkeley. He was a member of the Information Science and Technology advisory group at the Defense Advanced Research Projects Agency (DARPA) and is a member of the Defense Science Board. Prior to Intel, Manferdelli was a distinguished engineer at Microsoft and was an affiliate faculty member in computer science at the University of Washington. He was responsible for computer security, cryptography, and systems research, as well as research in quantum computing. At Microsoft, he also worked as a senior researcher, software architect, product unit manager, and general manager and was responsible for the development of the next-generation secure computing base technologies and the rights management capabilities currently integrated into Windows, for which he was the original architect. He joined Microsoft in February 1995 when it acquired his company, Natural Language Inc., based in Berkeley, California. At Natural Language, Manferdelli was the founder and, at various times, vice president of research and development (R&D) and chief executive officer (CEO). Other positions he has held include staff engineer at TRW Inc., computer scientist and mathematician at Lawrence Livermore National Laboratory, and principal investigator at Bell Labs. He was also an adjunct associate professor at the Stevens Institute of Technology. Manferdelli's professional interests include cryptography and cryptographic mathematics, combinatorial mathematics, operating systems, and computer security. He is also a licensed Radio Amateur (AI6IT). Manferdelli is a member of the National Academy of Engineering (NAE). He holds a bachelor's degree in physics from Cooper Union for the Advancement of Science and Art and a PhD in mathematics from the University of California, Berkeley.

HYRUM ANDERSON is the director of artificial intelligence (AI) and security at Cisco. Much of his career has been focused on defense and security, having directed research projects at the Massachusetts Institute of Technology (MIT) Lincoln Laboratory, Sandia National Laboratories, Mandiant, as the chief scientist at Endgame (acquired by Elastic), as the principal architect of Trustworthy Machine Learning at Microsoft, and CTO of Robust Intelligence (acquired by Cisco). Anderson co-founded the Conference on Applied Machine Learning in Information Security. He has authored more than 60 peer-reviewed academic publications and co-authored the book *Not with a Bug, But with a Sticker: Attacks on Machine Learning Systems and What to Do About Them*. He received his PhD in electrical engineering from the University of Washington, with an emphasis on signal processing and machine learning, and a BS and an MS in electrical engineering from Brigham Young University.

JOSIAH DYKSTRA is the director of strategic initiatives at Trail of Bits. He previously served for 19 years as a senior technical leader at the National Security Agency (NSA). Dykstra is an experienced cyber practitioner and researcher whose focus has included the psychology and economics of cybersecurity. He received the CyberCorps® Scholarship for Service (SFS) fellowship and is one of six people in the SFS Hall of Fame. In 2017, he received the Presidential Early Career Award for Scientists and Engineers from then President Barack Obama. Dykstra is a fellow of the American Academy of Forensic Sciences and a Distinguished Member of the Association for Computing Machinery (ACM). He is the author of numerous research papers, the book *Essential Cybersecurity Science* (2016), and co-author of *Cybersecurity Myths and Misconceptions* (2023). Dykstra holds a PhD in computer science from the University of Maryland, Baltimore County.

PAUL ENGLAND is a principal at Datica Research. Previously, he was a distinguished engineer and the manager of a team of researchers and engineers in Microsoft Research. England led or contributed to many of the computer industry's hardware-based security innovations over the past 20 years. Most notable is the field of trusted and confidential computing: a combination of novel cryptographic operations together with hardware and software environments for secure computation. Trusted computing primitives are now a feature of most mobile, client, server, and cloud computer systems, and the field remains an area of active research. England also contributed to the design of the first trusted platform module and led the team that developed the current version. He became interested in cyber-resilient systems through his work with National Institute of Standards and Technology (NIST) in developing NIST SP 800-193—Platform Firmware Resiliency Guidelines. Based on this, he subsequently worked with hardware partners and standards groups to develop architectures and hardware and software building blocks to enable secure and high-assurance recovery of devices that have been compromised by malware or misconfiguration. England is a member of the NAE. He received his PhD in condensed matter physics from Imperial College, London.

MARITZA JOHNSON is an expert on human-centered security and privacy with industry, teaching, and research experience. She is currently a principal at Good Research. In prior roles, Johnson was the founding director of the Center for Digital Civil Society at the University of San Diego, a user experience researcher at Google Research, and a technical privacy manager at Facebook. She is also an advisor to Confidencial, Inc. In 2011, her paper "The Failure of Online Social Network Privacy Settings" won the Future of Privacy Forum's Privacy Papers for Policy Makers Award. Johnson received an MS and a PhD in computer science from Columbia University in 2008 and 2012, respectively, and a BA from the University of San Diego in 2005.

ANGELOS D. KEROMYTIS is the John H. Weitnauer Endowed Chair Professor and Georgia Research Alliance Eminent Scholar at the Georgia Institute of Technology (Georgia Tech). His field of research is systems and network security and applied cryptography. He joined Georgia Tech from DARPA, where he served as program manager in the Information Innovation Office (I2O) from 2014 to 2018. During that time, he initiated five major research initiatives in cybersecurity, managed a portfolio of nine programs, and supervised technology transitions and partnerships with numerous elements of the Department of Defense (DoD), the Intelligence Community, law enforcement, and other parts of the U.S. government. For his work, he received the DAPRA Superior Public Service Medal. Before DARPA, he served as the program director with the Computer and Network Systems Division in the Directorate for Computer and Information Science & Engineering at the National Science Foundation, where he co-managed the Secure and Trustworthy Cyberspace program and helped initiate a number of cross-disciplinary and public–private programs. Prior to his public service tour, Keromytis was a faculty member of the Department of Computer Science at Columbia University, where he founded the Network Security Lab. He is an elected fellow of ACM and the Institute of Electrical and Electronics Engineers (IEEE). He has 63 issued U.S. patents and more than 250 refereed publications. His work has been cited more than 30,000 times, with an h-index of 86 and i10-index of 262. He has founded four new technology ventures and is currently serving as the president for two of them. He is a certified PADI Master Instructor, with more than 800 dives. He received his PhD (2001) and MSc (1997) in computer science from the University of Pennsylvania and his BSc in computer science from the University of Crete, Greece.

WENDY NATHER is the senior research initiatives director at 1Password. She was previously the director of advisory chief information security officers (CISOs) at Duo Security and the research director at the Retail ISAC, where she was responsible for advancing the state of resources and knowledge to help organizations defend their infrastructure from attackers. Nather was also the research director of the Information Security Practice at independent analyst firm 451 Research, covering the security industry in areas such as application security, threat intelligence, security services, and other emerging technologies. Nather has served as a CISO in both the private and public sectors. She led information technology security for the EMEA region of the investment banking division of the Swiss Bank Corporation (now UBS), as well as for the Texas Education Agency. She is the co-author of *The Cloud Security Rules: Technology Is Your Friend. And Enemy* and Splunk's *Bluenomicon: The Network Defender's Compendium*. She was inducted into the Infosecurity Europe Hall of Fame in 2021. She serves on the board of directors for Sightline Security, an organization that helps provide free security assessment services to nonprofit groups. Nather is a senior fellow at the Atlantic Council's Cyber Statecraft Initiative and a steering committee member for the IST Ransomware Task Force.

STEFAN SAVAGE is the Irwin and Joan Jacobs Professor of Information and Computer Science at the University of California, San Diego (UCSD). He currently serves as the co-director for UCSD's Center for Network Systems and as a founding member of the school's Center for Healthcare Cybersecurity. Savage is known for his work on network security and reliability, cybercrime economics and defense, and the empirical measurement of cybersecurity and cyberinfrastructure. He is a member of the NAE and the American Academy of Arts and Sciences, a MacArthur fellow, an ACM fellow, and is the recipient of ACM's Prize in Computing and the American Association for the Advancement of Science's Golden Goose award. He received his PhD in computer science and engineering from the University of Washington and a BS in applied history from Carnegie Mellon University (CMU).

WILLIAM L. SCHERLIS is a professor of computer science at CMU and the special advisor to the Software Engineering Institute, a DoD federally funded R&D center at CMU. He recently served as the director of DARPA's I2O, leading program managers in the development of programs in cybersecurity, artificial intelligence, secure software, and information operations. At CMU, he served for more than a decade as head of the Software and Societal Systems Department, which hosts research and educational programs related to software development, security and privacy, Internet of Things and mobility, AI engineering, social network analysis, and related topics. He founded the CMU PhD program in software engineering and led it for its first decade. His research relates to software assurance, cybersecurity, software analysis, and assured safe concurrency. He has led several large research projects, including the CMU NSA Science of Security Lablet and the CMU/NASA High Dependability Computing Project. He served as the program chair for technical conferences, including ACM Foundations of Software Engineering and ACM Partial Evaluation and Program Manipulation. Scherlis has led multiple national studies including the National Research Council study that in 2010 produced the report *Critical Code: Software Producibility for Defense*. He has testified before Congress on the AI workforce, federal software sustainment, computing technology and innovation, and on roles for a federal chief information officer. He is a Life Fellow of IEEE and a Lifetime National Associate of the National Academy of Sciences. Scherlis received an AB magna cum laude from Harvard University in applied mathematics and a PhD in computer science from Stanford University, with an intervening year in the Department of Artificial Intelligence at the University of Edinburgh as a John Knox fellow.

MARK SEIDEN currently serves as the security advisor to the Internet Archive. He previously held the position of associate in computer science at Columbia University and developed and taught a master's level operating system security course for the University of California (UC), Berkeley, School of Information. In addition, he is engaged in a DARPA-funded research project with UC Santa Cruz and has offered his expertise in more than 50 criminal and civil cases. With a programming career that spans since the 1960s, Seiden has collaborated with diverse companies and research institutions, making significant contributions to software engineering, network technologies, operating systems, and physical security. In recent years, Seiden served for 15 years in the ICANN Security and Stability Advisory Committee and actively participated in multiple National Academies' studies addressing technological risk. His noteworthy affiliations include roles at IBM Research, Lucasfilm, Yahoo, Xerox PARC, Bell Labs, Bellcore, and IRCAM. Seiden earned his SM in computer science and electrical engineering from Columbia University in 1981.

WINDOW SNYDER is a security industry pioneer and the CEO and founder of Thistle Technologies. Snyder is the former chief security officer at Square and Fastly. She previously spent 5 years at Apple responsible for security and privacy strategy and features for OS X and iOS. Other roles include chief software security officer at Intel, chief security something-or-other at Mozilla, and a founder at Matasano, a security services and product company based in New York. Snyder is the co-author of *Threat Modeling*, a manual for security architecture analysis in software.

MARY ELLEN ZURKO is a technical staff member at the MIT Lincoln Laboratory. She has worked in product development, early product prototyping, and research and has more than 20 patents. She defined the field of user-centered security in 1996 and has worked in cybersecurity for more than 35 years. She was the security architect of one of IBM's earliest clouds. She was a founding member of the National Academies' Forum on Cyber Resilience and serves as a Distinguished Expert for NSA's Best Scientific Cybersecurity Research Paper competition. Her research interests include unusable security for attackers, zero trust architectures for government systems, security development and code security, authorization policies, high-assurance virtual machine monitors, the web, and PKI. Zurko received an SB and an SM in computer science from MIT. She has been the only "Mary Ellen Zurko" on the web for more than 25 years.

D
Glossary

An **attack surface** consists of the set of all possible access pathways through which an attacker could engage with a cyber system and its operations.

Cyber refers to the technology and culture of computers, information technology, and communications. Cyber encompasses all technical artifacts that incorporate programmable digital devices (as opposed to analog and fixed function devices). There are many related definitions, for example, see *Cybersecurity Myths and Misconceptions*.[1]

Cyber-enabled refers to devices and systems that incorporate or embed cyber technology.

Cyber-physical system (CPS) is a system that incorporates programmable logic, sensors, and actuators that enable it to perceive and engage with the "real world." CPS are employed in industrial controls (control systems), domestic devices (cameras, digital thermostats), network control (routers and switches), and infrastructure operations. Modern cars are examples of CPS and contain many subordinate elements that are themselves CPS. Most modern buildings are also examples, using diverse sensors and actuators for HVAC, elevators, lighting, security, and the like.

Cyber resilience is the ability of an organization, system, or function to withstand and recover from cyberattacks.

---

[1] E.H. Spafford, L. Metcalf, and J. Dysktra, 2023, *Cybersecurity Myths and Misconceptions*, Pearson.

Cybersecurity is the use of technologies and practices to protect systems, networks, and data from cyberattacks.

Internet of Things (IoT) refers to "small" cyber-physical systems (CPS) that are connected on a network, and often also to the public Internet. Because most CPS are network connected, the distinction between operational technology (OT) and IoT, if there is one, centers around the "smallness" of the related system. OT is used to distinguish between operational systems like automated machinery and conventional "IT" systems that support business operations. Needless to say, the distinction is increasingly muddled.

Large language model (LLM) is a type of neural-network-based artificial intelligence model that uses deep learning to perform natural language processing tasks, typically by predicting, in response to a prompt, the next tokens (roughly, words) in a stream of natural language. The models are fundamentally statistical predictors that are trained on large amounts of data to learn patterns and rules of language.

Least privilege is a principle of systems design and operation whereby those who perform tasks, whether they be humans or other system elements, are granted access and other privileges just sufficient to perform the intended task. It is analogous to the security concept of "need to know."

Memory safety, similar to type safety, is a property of some programming languages. Memory safety means that memory cannot be accessed except according to the identified rules. These rules generally include prohibiting access to regions of memory outside of the regions or defined bounds of storage associated with a particular computational process. Memory safety also includes prohibiting access to memory that is not currently allocated to specific objects—memory not in use cannot be referenced. Memory safety can also include mechanisms to regulate access to memory that might be exposed to multiple threads of execution, for example to prevent data races where one thread sees objects that are currently being modified by another thread. Memory safety and type safety can dramatically reduce vulnerabilities that enable buffer overflow attacks and use-after-free attacks. Go, Rust, and Java are examples of languages that are both type safe and memory safe. C is decidedly not memory or type safe. In C, for example, the bit sequences in memory are always exposed, and can be interpreted as many kinds of objects and modified without regard to those interpretations.

NPM stands for node package manager. NPM is a library and registry for more than 1 million JavaScript software packages used in the development and operation of web applications.

Reliability refers to the quality of being trustworthy or of performing consistently well. See also the definition of *resilience* and *trustworthiness* below.

Resilience: The classic definition is "the capacity to withstand or to recover quickly from difficulties." There is an entire discipline of resilient engineering that evolved with a focus on critical infrastructure like bridges, highways, municipal water and sewer systems, and electrical distribution systems. In these cases, there are traditional core resilience metrics such as availability and recovery time (https://in.nau.edu/comptroller/bcdr-glossary has an extensive list of related definitions). For cyber systems, resilience more often refers to the capacity of a system to continue to operate in an effective manner when there are internal faults and attacks, though perhaps in a degraded manner. Faults can include human errors by operators and users. The military uses a term "operate through" to refer to this important characteristic. When multiple systems are interconnected, resilience includes avoidance of cascading failures and other amplifying effects, instead localizing and when necessary isolating portions of the overall system that have been impaired or compromised. A challenge for cyber systems is that they often do not have clearly measurable qualities—because vulnerabilities and faulty operations are often not visible and even unknown. In common usage, a resilient cyber system can repel or otherwise protect against attacks whose effects may not be visible thus incorporating the notion of trustworthiness inherent in reliability.

Root of Trust (RoT) refers to a small "secure" component that measures aspects of software and firmware content in order to assure integrity of configurations. The RoT component typically contains long-term, critical secrets that allow a system to attest to safety properties. RoT is important, for example, in ensuring a provider of confidential information that a remote system will only operate on that information in a permitted manner.

Safe is the condition of being protected from or unlikely to cause danger, risk, or injury, including to human operators, users, and others.

Secure is the state of being free from danger or threat, and operating in a trustworthy manner (see *reliability*). In the context of cyber, secure systems are resistant to attack.

Software bill of materials (SBOM) is an inventory that identifies the software elements that are assembled into an integrated software application, including components, libraries, and services. These can include components and libraries from vendors, custom developers, and open-source projects, as well as services from public and private cloud providers of various kinds, including commercial software as a service. The SBOM inventory can be structured hierarchically to represent how system elements are recursively composed from subordinate system elements. It is sometimes appropriate to interpret the scope of SBOM more broadly, to also include elements such as architectural design patterns, code snippets copied from example code, and other elements that may not have identified producers.

A **system** is a set of components and services operating as an ensemble as parts of a mechanism or in an interconnecting network. Examples range from software systems to a railroad system.

Trust is a judgment made by people regarding the reliability or cybersecurity risk associated with a system. The best basis for trust is trustworthiness of the system. But trust can also be derived from reputation, alignment of business incentives, and other extrinsic attributes.

Trusted computing base (TCB) refers to the collection of software and hardware technology that is most critical to secure and reliable operations of a system. A TCB is intended to be the smallest portion of a system for which operations must be completely trustworthy and reliable. Elements of a system not in the TCB might possibly misbehave but without catastrophic consequence.

Trustworthiness is an attribute of a system or other artifact that relates to its reliability and security. Judgments of trustworthiness are derived from intrinsic attributes of a system and its design, with judgments generally based on evidence.

Type safety is a property of some programming language designs where there is enforcement of the manner in which lower-level representations, such as a sequence of bits, are interpreted as higher level abstract objects, such as sensor signals, data from a table, or imagery. The various kinds of objects are referred to as *types* or *abstractions*, and given names. Type safety additionally assures that low-level representations cannot be interpreted or tampered with except according to the rules associated with the higher-level types.

Vulnerabilities are exposed features of attack surface that could admit the possibility of a successful attack, or exploit. Attack surface could be exposed network portals, web application programming interfaces (APIs) exposed to browsers, software APIs exposed to rogue software elements, physical components attached to a shared hardware bus, and so on. Attack surfaces can also include physical elements of hardware, such as for timing and power attacks. The span or extent of attack surface can be difficult to quantify. In larger systems, some attack surfaces offer more opportunity for an attacker than others, and it is a security design principle to minimize these "high consequence" attack surfaces (see *trusted computing base*).

Work factor refers to the extent of human attention and computational resources to perform an identified attack. It is often used in an adversarial context such as the "extent of computational effort to break a cipher without the key." It is also used to describe the extent of difficulty presented to an attacker to successfully penetrate a vulnerable attack surface. For example, there may be less work factor in an attack launched over a network than in an attack that requires a human agent to be physically proximate to the target.

Zero trust (ZT) is a design methodology for interconnected systems and for larger systems with multiple subordinate subsystems that emphasizes the principle of least privilege supported by identity and authorization techniques whereby credentials must be reasserted when internal system and operational boundaries are crossed. ZT is designed to avoid Maginot-Line situations where an agent (e.g., a human operator or malware code), having penetrated a perimeter, then has unrestricted access to everything contained. Instead, there are repeated challenges for identity and authorization that have the effect of creating resilience in operations, as described above.